BIG DATA'S THREAT TO LIBERTY

BIG DATA'S THREAT TO LIBERTY

Surveillance, Nudging, and the Curation of Information

HENRIK SKAUG SÆTRA
Faculty of Computer Sciences, Engineering and Economics,
Østfold University College, Halden, Østfold, Norway

Academic Press is an imprint of Elsevier
125 London Wall, London EC2Y 5AS, United Kingdom
525 B Street, Suite 1650, San Diego, CA 92101, United States
50 Hampshire Street, 5th Floor, Cambridge, MA 02139, United States
The Boulevard, Langford Lane, Kidlington, Oxford OX5 1GB, United Kingdom

Copyright © 2021 Elsevier Inc. All rights reserved.

No part of this publication may be reproduced or transmitted in any form or by any means, electronic or mechanical, including photocopying, recording, or any information storage and retrieval system, without permission in writing from the publisher. Details on how to seek permission, further information about the Publisher's permissions policies and our arrangements with organizations such as the Copyright Clearance Center and the Copyright Licensing Agency, can be found at our website: www.elsevier.com/permissions.

This book and the individual contributions contained in it are protected under copyright by the Publisher (other than as may be noted herein).

Notices

Knowledge and best practice in this field are constantly changing. As new research and experience broaden our understanding, changes in research methods, professional practices, or medical treatment may become necessary.

Practitioners and researchers must always rely on their own experience and knowledge in evaluating and using any information, methods, compounds, or experiments described herein. In using such information or methods they should be mindful of their own safety and the safety of others, including parties for whom they have a professional responsibility.

To the fullest extent of the law, neither the Publisher nor the authors, contributors, or editors, assume any liability for any injury and/or damage to persons or property as a matter of products liability, negligence or otherwise, or from any use or operation of any methods, products, instructions, or ideas contained in the material herein.

Library of Congress Cataloging-in-Publication Data
A catalog record for this book is available from the Library of Congress

British Library Cataloguing-in-Publication Data
A catalogue record for this book is available from the British Library

ISBN: 978-0-12-823806-6

For information on all Academic Press publications visit our website at
https://www.elsevier.com/books-and-journals

Publisher: Mara Conner
Acquisitions Editor: Chris Katsaropoulos
Editorial Project Manager: Rafael G. Trombaco
Production Project Manager: Omer Mukthar
Cover Designer: Miles Hitchen

Typeset by TNQ Technologies

Working together
to grow libraries in
developing countries

www.elsevier.com • www.bookaid.org

Contents

Foreword vii

Preface xi

Chapter 1 Introduction 1

Introduction 1

The boundaries of the current undertaking.......................... 3

Philosophical, theoretical, and methodological foundations 5

Who is this book for? 8

The structure of the book 10

References 12

Chapter 2 Technologies and society 15

Introduction 15

Big Data as a logic and a system of technologies 15

Liberty in its various guises 23

References 30

Chapter 3 Liberty under surveillance 35

Introduction 35

Big Data surveillance 35

When Big Brother sees you 38

Freedom under the gaze of Big Data 46

Conclusion.. 48

References 49

Chapter 4 Big Data nudging and liberty 51

Introduction 51

Nudging 52

When the nudge is powered by Big Data............................... 54

Nudging and liberty 58

vi Contents

When nudge comes to shove 74

Conclusion...76

References...77

Chapter 5 The algorithmic tyranny of perceived opinion 79

Introduction 79

Big Data and information 80

Freedom 87

The threat of a tyranny of perceived opinion.....................................92

Conclusion..95

References...96

Chapter 6 The three threats in concert 99

Introduction 99

Nudging by curation of information... 100

Privacy is a public good 105

Summary 110

References... 110

Chapter 7 Liberty in the era of Big Data...113

Introduction 113

Preliminaries 114

Interference: beyond the physical 122

Interference and privacy 126

The implications for liberal policy 128

Summary 130

References... 131

Chapter 8 Conclusion 133

References... 136

Index 137

Foreword

A few months ago, I purchased an Amazon Echo. If you are not familiar with it, the Echo is a smart speaker with an integrated cloud-based AI program called Alexa. The device now sits on my kitchen counter, silently blinking in blue, waiting patiently for my commands. "Alexa, what's the weather like today?," "Alexa, give me my news update, please!," "Alexa, remind me to call my mother at 3pm!."

This was not an impulse buy; I purchased it for a reason. Like Henrik Skaug Sætra, I am a philosopher and ethicist of technology. My general feeling is that if I am going to write about technology, it is incumbent on me to actually use it and be familiar with its features. But, given my day job, I am aware of the risks that such devices pose. They are, after all, surveillance machines. You welcome them into your home and then they track your voice commands, learn your preferences, and feed Amazon's databases with information they can use to hone their products and services. That said, Alexa has a number of conveniences that I like. It is great for setting reminders and keeping track of your calendar. Its AI is also reasonably impressive, capable of answering factual questions and performing complex calculations. My inner child rejoices every time I ask it to do something. I feel as if I have stepped onto a science fiction movie set, speaking to a computer as if it were a normal and mundane part of my life (which, in a sense, it now is).

Recently, however, I have noticed a change in my behavior around Alexa. I am always conscious that it might be listening. I tell my wife to watch what she says in its vicinity. Although Alexa is only supposed to pay attention when you use its "wake word" ("Alexa" in this instance), a quick review of the voice history suggests that it can make mistakes. It sometimes listens and records information before deleting it as irrelevant. I still like the movie and book recommendations that Alexa gives me, but, despite the inconveniences this causes, I now tend to switch it off whenever I am not actively using it.

Here's the question: Does a technology like Alexa undermine my liberty? In a sense, it feels like it does. If I invited another person into my home, to observe my every move, I would undoubtedly act differently as a result. I would be more guarded in what I say; more cautious in what I do. I would start putting on a show rather than being my true self. That sounds like the opposite of being free. On the other hand, Alexa does allow me to outsource some of my cognitive burden. I do not have to keep track of dates and appointments anymore. Alexa will remind me of these when needed. This frees me up to think about other things that are more important to me. Similarly, Alexa gives me more choices than I had before when it comes to music, radio, and other forms of information. More choices and more time sound like a good thing when it comes to protecting liberty. It seems like there are arguments to be made on both sides. The technology both undermines and promotes my liberty.

Or does it? If you, like me, are confused about this issue, then you will be glad that you picked up this book. In *Big Data's Threat to Liberty*, Henrik Skaug Sætra expertly guides the reader through the thicket of concepts and issues that lie at the heart of my confusion about Alexa. Liberty, it turns out, is a complex idea. People often claim that technology undermines or promotes liberty, but they often do not clarify what they mean by this. It is assumed that we know what liberty is. But liberty has many faces. There is negative liberty (freedom from interference and domination) and positive liberty (the power to be who you want to be). There are also perfectionist and non-perfectionist forms of both, as well as legitimate and illegitimate forms of interference with liberty. How are we to make sense of this complexity?

Sætra has the answers. In a crystal clear exposition of big data and the threat to liberty, Sætra encourages us to avoid the danger of talking past each other. He clarifies what is meant by "liberty" and what is meant by "big data" and then, in a series of rigorously formulated propositions and arguments, he identifies three core threats that big data poses to liberty: the surveillance threat, the nudging/shoving threat, and the information curation threat. Building on this, he then explains how these threats are linked to power relations in society and why it is important for us to care about them.

The book he has produced speaks for itself. I cannot do a better job summarizing its key claims than he has already done (indeed, one of the virtues of the book is its regular summaries of key arguments). I would, however, like to comment on two features of the text that I think are worth highlighting at the outset.

First, although this book is accessible to anyone with an interest in technology and its impact on society, philosophers and ethicists of technology may find it particularly useful as an embodiment of the merits of an analytical approach. Sætra does not shy away from empirical data when it is relevant to his arguments, but his study is largely a conceptual one. He clarifies concepts and definitions and then uses these to build logically valid arguments. This is the classic methodology of political philosophy, but it is one that many people now feel slightly embarrassed about. To sit in one's armchair and pontificate about the nature of liberty and power is often devalued and seen as an incomplete form of scholarship. Sætra's book should reassure anyone who feels embarrassed about this approach. It demonstrates the unambiguous value of analytical rigour and clarity. Indeed, I would argue that Sætra's book is far more illuminating than many similar books with more complex methodologies.

Second, Sætra's book is unabashedly liberal in its outlook, engaging with the best available liberal political philosophy and using it to scrutinize the impact of big data on our lives. "Liberalism," sadly, has become a dirty word in some parts of the modern world. It is often associated with a shallow, overly atomistic, and individualistic outlook on human life. Liberals, we are told, are obsessed with the state as an institution that can undermine our liberty, but less concerned about private corporations and social groups that can do the same. A more nuanced, post-liberal perspective is sometimes urged in response to these perceived shortcomings. But Sætra shows that many of these criticisms are misguided. There is plenty of life in liberalism yet. Liberalism is a complex, multifaceted normative theory that can account for different forms and sources of power. Sætra uses liberal theory to great effect throughout this book to shed light on big data. I, for one, learned a lot from reading this book.

Turn the page and you will too.

John Danaher,
NUI Galway,
Ireland

Preface

The journey from an immature idea to a finished book can be a strange one, as the journey of the idea that became this book thoroughly demonstrates. It is the direct result of my pursuit of a Ph.D., but it originated at a destination far removed from Big Data and Artificial Intelligence. My interests have always been varied, and I started the Ph.D. process working on environmental ethics and the classical social contract tradition. There is one common denominator between where I started and where I ended up, however—political philosophy. However, I have also always been interested in and intrigued by technology. When the idea emerged of combining technology with my love for philosophy, the proverbial pieces fell in place.

My ideas have a tendency to be born in a format not containable by the bounds available to pursue them. This was certainly the case with my desire to describe the threats posed by Big Data. When I decided to first test the waters where technology is combined with political philosophy, I wrote a draft article titled "Freedom under the Gaze of Big Brother: A Liberal Defence against the Perils of Big Data." It was an article bursting at its seams, as I wanted to make a number of what I perceived to be important points at once. Big Data, I argued, was problematic due to its effects on liberty through surveillance, nudging, and the curation of information. My Ph.D. supervisors got this draft of about 13,000 words, and proceeded to calmly relay the message that this is not really *one* of the articles in a dissertation. It is the beginning of several articles, and in and of itself it attempts to do far too much. The draft was submitted on September 18, 2018. Fast forward 2.5 years and that draft has evolved into my dissertation and, later on, this book.

My dissertation, defended at the University of Oslo, Norway, consisted of three articles on three distinct threats to liberty posed by Big Data. This required me to cut apart the three ideas contained in my original draft, while I was able to write an introduction that explained how they were related. However, as I originally intended to write a coherent story describing the threat of Big Data, I reached out to Elsevier, who had published all the three articles, with a proposal for this very book, where I tied things back together in a format more conducive to telling a more comprehensive story. After defending my dissertation in the autumn of 2020, I then proceeded to reintegrate the various arguments into a book describing its totality.

Chapters 3–5 contain the majority of the articles of my dissertation, namely "Freedom under the Gaze of Big Brother: Preparing the Grounds for a Liberal Defence of Privacy in the Era of Big Data" (Sætra, 2019a), "When Nudge Comes to Shove: Liberty and Nudging in the Era of Big Data" (Sætra, 2019c), and "The tyranny of Perceived Opinion: Freedom and Information in the Era of Big Data" (Sætra, 2019b). The latter article has largely become Chapter 5, but parts of it—another example of me originally trying to do too much at once—is now further developed and has become part of Chapter 6. Remnants and parts of the introductory chapter to my dissertation can be found throughout the book, and particularly in Chapters 1, 2, and 6. While writing the articles that become my dissertation, however, I also started developing new ideas I wanted to pursue, and "Privacy as an Aggregate Public Good" (Sætra, 2020) was a direct result of the ideas that emerged from the writing about surveillance and privacy. Edited excerpts from this article have also become part of Chapter 6, where it is used to further develop the importance and nature of privacy as it relates to the described threats. While much in this book thus stems from the articles I have published, much is also new, updated, and further developed. The seventh chapter was one I had originally wanted to include in my dissertation, but once again wiser voices urged me to focus my attention on a manageable topic. It was thus excluded from the dissertation, and I am very happy to be able to present in this book the complete argument that I once vaguely saw, and which was developed through a number of interested detours and partial efforts.

It is sometimes said that no man is an island. This is certainly true when it comes to writing a dissertation. When I started working on these topics, I could not imagine that this is where I would end up, adrift, and on the move, so perhaps more like a ship than an island, then. But never alone or without guidance, which I suppose is what the saying is trying to get at. I want to extend my deepest appreciation and gratitude for the support and guidance I have received from my supervisors, Prof. Raino Malnes and Prof. Knut Midgaard. When I started contemplating doing a Ph.D., I had no doubt that these were the two I wanted by my side, and I was fortunate enough for that to happen. They even stuck with me through the various ideas and changes I proposed—pulling me back to constructive paths when they were bad and supporting the better ones.

I also wish to thank the fellow ships I have happened to come in contact with through my journeys on these academic waters; Harald Borgebund, Stuart Mills, and Eduard Fosch-Villaronga in particular. I have also found fellow seafarers from all over the world, and some of them have read my manuscripts and provided invaluable feedback through the peer-review processes involved in the publication of my articles. I do not know who you are, but thank you for taking the time to be both constructive and helpful. An island I am not, and if I am a ship it is not a solitary one. My family is always there for me, no matter what, and I could not begin to recount the instances where this has been proven. Thank you, for all your love and support. I wish to extend a special thank you to my parents, my son Brage, and my wife, Christine.

References

Sætra, H. S. (2019a). Freedom under the gaze of Big Brother: Preparing the grounds for a liberal defence of privacy in the era of Big Data. *Technology in Society, 58*, 101160.

Sætra, H. S. (2019b). The tyranny of perceived opinion: Freedom and information in the era of big data. *Technology in Society, 59*, 101155.

Sætra, H. S. (2019c). When nudge comes to shove: Liberty and nudging in the era of big data. *Technology in Society, 59*, 101130.

Sætra, H. S. (2020). Privacy as an aggregate public good. *Technology in Society, 63*, 101422. https://doi.org/10.1016/j.techsoc.2020.101422.

Introduction

Introduction

Big Data permeates just about all aspects of modern life. We encounter it at home and at work, when we work out and when we relax. From old to young—even the unborn—are tracked and analyzed, and in turn influenced by Big Data in a variety of ways. Much of this goes unnoticed by most of us, and while some rejoice at the new opportunities provided by new technologies, others are deeply concerned about how our lives and relations are affected by ubiquitous surveillance and the creation of Big Data-based personality profiles that are used to guide us toward the often-unknown ends of others.

But who are right—the optimists or the skeptical ones? Both are. I fully accept the many benefits of Big Data, as discussed below, but this book mainly focuses on one possible explanation of why the pessimists have good reason to be wary. This analysis is both important and necessary, because the debates about the dangers of Big Data are riddled with a wide variety of opinions on what the nature of the dangers are, and too little attention has been paid to the analysis of how *liberty* is threatened by surveillance, Big Data-based nudging, and the algorithmic curation of data. These threats are thus undertheorized and arguably of great importance in today's world.

Granted, many *mention* the word liberty—or freedom (the two terms are here used interchangeably)—but too few take the time to define and explain what they mean by this concept (Yeung, 2017; Zuboff, 2019). The main puzzle that led to the writing of this book is that the threats to liberty that we at times intuitively perceive are usually either (a) lost in vagueness due to the concept of *liberty* being employed without a clear understanding of its content, or (b) lost in definitions if we employ flawed traditional understandings of what liberty really is. When the threat in question is posed by an all-encompassing phenomenon such as Big Data, we have both an interesting and potentially very important puzzle on our hands. The main research question I seek to answer

Big Data's Threat to Liberty. https://doi.org/10.1016/B978-0-12-823806-6.00008-9
Copyright © 2021 Elsevier Inc. All rights reserved.

is: Does Big Data threaten liberty, and if so, what are the most important ways in which it does so?

The phenomena I discuss are related to how we are constantly under surveillance by various actors. Data is gathered from our houses, cars, smartphones, various devices at home, and through social networks, commercial and public websites and services, in addition to general surveillance through ubiquitous camera surveillance combined with facial recognition (Crawford et al., 2019). For example, the average Londoner (the UK has the highest concentration of surveillance cameras per capita in the West) is by some estimated to be captured on private and public security cameras about *300* times each day (Dormehl, 2019; Olson, 2019). Is being constantly watched compatible with liberty?

Furthermore, the information gathered is used to influence our actions. Detailed personality profiles are employed in order to make us purchase products and services, vote, click on news stories, pay taxes, and the list goes on. This happens through tailormade *nudges*—techniques of influence aimed at irrational and subconscious mechanisms. These nudges are now created and delivered with a level of precision only possible with Big Data-driven algorithmic curation of data. Are we free when manipulated, or *coerced*, in this manner?

Finally, the information we receive in various media is curated by algorithms, and even *people* are curated in manners meant to satisfy our desires. By providing us with what the algorithm finds it likely that we want, we are spared from exposure to unpleasant information, and even unpleasant people. The ideological landscapes we traverse are in turn potentially increasingly characterized by conformity, and a concomitant tyranny of popular opinion becomes ever more coercive as this occurs. These are the three main threats I explore in this book, with the aim of showing the various ways in which they are threats to liberty.

Technological change has always been an important part of human existence. In fact, the industrial revolutions—some of the most important life and society changing events—are the results of such change. The first industrial revolution was based on the shift from animate to inanimate power sources, while the second followed the development of more efficient ways to *transfer* power in electric and internal combustion engines (Barley, 2020). The third followed the development of the technologies that allowed the creation of computers (Barley, 2020). Today, some argue that technologies such as AI and new biotechnologies form the basis of a *fourth* industrial revolution (Schwab, 2017). Barley (2020), however, convincingly argues in favour of seeing both the third and the so-called fourth revolution as two phases of the *control* revolution.

No matter what we call the macro level results of new technologies, and whether or not we consider such changes beneficial and unavoidable, we must exert great effort to analyze their implications. Because when we understand the technologies in question, we are in a position to both understand and direct the changes that accompany them. If new technologies, here Big Data, have implications we do not like, there is a better chance of countering them. While technologies have a certain form of power to shape our societies, I reject the technological determinism that at times surfaces when technology is discussed (Barley, 2020; Heilbroner, 1994; Marx & Smith, 1994; Wyatt, 2008).

One problem with technology is that it does not always do only what we desire. There are "unanticipated social consequences" and such consequences are the ones I here describe (Collingridge, 1980, p. 16). According to *Collingridge's dilemma*, technology can be shaped when it is new, but at that point we do not fully understand its implications. When technology has matured, the implications are known, but changing it becomes quite difficult. Big Data is now relatively mature, and thus Collingridge's dilemma implies that this technology might be hard to control. The harmful effects, one of which is lost liberty, can now be seen, and we must do what we can to regain control.

This is one of the puzzles that motivated this book: How have we allowed these technologies to become so ubiquitous, when their consequences are, in the respects I consider, potentially quite undesirable? Part of the answer, I argue, is that we have not fully understood how these technologies affect our liberty.

Before we can gain control, we must understand, and this is where this book is meant to contribute. Bruno Latour (1999, p. 304) discusses blackboxing, and "the way scientific and technical work is made invisible by its own success" Technologies can become entirely opaque due to their ubiquitous nature, but philosophical analysis of the concepts involved can increase the transparency of the box in question.[1]

The boundaries of the current undertaking

This book is a rather focused one, and I make no attempt to deal fully with all aspects and implications of Big Data. A short

---

[1]Another issue is related to the lack of transparency and explainability when Big Data and artificial intelligence (AI) works its magic, and how those in charge of systems built on these technologies may have certain incentives to make sure that the box *stays* black (Sætra, 2021b).

description of what I will *not* do—that others have done—is thus in order. Firstly, I have not attempted to give a full description of the various *positive* impacts of technology. Governments, businesses, citizens, and consumers alike have good reasons to like Big Data. Businesses can target consumers better, which leads to more effective marketing and the ability to charge more for services and products that satisfy the customer's preferences (Bello-Orgaz, Jung, & Camacho, 2016; Chen, Chiang, & Storey, 2012; Cohen, 2012). With enough information about an individual, businesses often know more about individuals than even their close friends and relatives (Youyou, Kosinski, & Stillwell, 2015). In addition, services such as insurance can be priced on the basis of more and better data on risks and behavior (Baruh & Popescu, 2017). Automation and cost reduction in production is a related advantage, but more due to AI than Big Data—terms I explain in more detail in the next chapter. For consumers, increased satisfaction of preferences is a boon, and services such as social networks may by some be seen as improving people's lives. Big Data also leads to new possibilities for scientific progress in all sciences, including those leading to health benefits (Chen et al., 2012). In addition, politicians can use Big Data for opinion mining and social network analysis, while social media and Big Data analytics can support political processes and accountability (Chen et al., 2012). While surveillance is controversial, there are obvious security benefits to be reaped from Big Data, in terms of crime prevention, epidemic intelligence, etc. (Bello-Orgaz et al., 2016; Chen et al., 2012; Sætra, 2020). In addition, attention is increasingly being paid to how AI and Big Data influence the UN's sustainable development goals (Sætra, 2021a; Vinuesa et al., 2020).

Secondly, I have not attempted to conclude on the question of what should be given most weight when liberty collides with utility, or security. A related issue is that of *fairness*, especially when discussing the nonneutrality of algorithms and the biases inherent in them (Buolamwini & Gebru, 2018; Köchling & Wehner, 2020; Noble, 2018). What, for example, do we say when insurance companies increase the premiums for those most in need of insurance, or when new technologies are more—or less— reliable, and create different results, for different groups of individuals? Julie Cohen (2012, p. 1931) states that "a liberal democracy cannot simply deploy surveillance technologies to close the gap unfulfilled and unfulfillable by perfect technologies of justice", Reuben Binns (2018) provides an account of fairness in machine learning based on political philosophy, and Kate Crawford et al. (2019) give a detailed account of the current state of bias in artificial intelligence and efforts to regulate it. I will also note that in terms of security and politics, Big Data has also led to *problems*, such as the Cambridge Analytica case in 2018 and Russia's alleged use of social media as an instrument for influencing foreign policy (Greenfield, 2018; McKew, 2018).

Philosophical, theoretical, and methodological foundations

The importance of technology has not been lost on *any* academic discipline, and a wide array of academics from different disciplines are today approaching the questions I discuss from different angles. Few topics are more interdisciplinary than the societal and individual effects of new technologies, but real and useful interchange between the disciplines involved requires us to acknowledge where we come from, what our frame of reference is, and how we choose to approach the topic. Myself being a political philosopher, I face a range of hurdles when I engage with the work of STS scholars, sociologists, computer scientists, etc.— *unless*, that is, they take the time to introduce and explain core concepts and theories in their work. In line with this, a few words regarding the philosophy, theories, and methodology here employed are in order, even if such an account cannot even begin to approach completeness. Hopefully it fosters transparency and openness, and hopefully others will take the time to do the same, in order to foster truly interdisciplinary debate on the societal and individual effects of technology. Those inclined to lose their patience with comprehensive philosophical and theoretical backgrounds can without much loss move straight to the next chapter at this point, while the rest can read on for a better understanding of where this book is grounded.

Mark Coeckelbergh (2018) laments the fact that much analysis of technology relies on *implicit* assumptions, and states that the field needs political philosophy, which offers "excellent resources for thinking about justice, equality, freedom, democracy, and other political principles which, unfortunately, are not often used by philosophers of technology to discuss the societal impact of technology" (p. 6). Explicit fundamental assumptions are important, and this book is an attempt to discuss technology by way of political theory, with a particular focus on a clear conception of what sort of *freedom* I discuss.

Political science is often quite empirical, but there is also a normative branch—political philosophy or normative political theory (Dowding, 2015)—in which this book belongs. This branch is further split into a variety of approaches, and my approach is most accurately described as analytical philosophy (AP), which is now considered the dominant tradition in the English-speaking world of philosophy (Beaney, 2013). AP is not easily defined, however, as it is less of a unitary doctrine than a "loose concatenation of approaches to problems" (Stroll, 2000, p. 5). There is a wide array of approaches that is now categorized as *analytical philosophy*, but one aspect that unites this motley crew is the "implicit respect for argument and clarity" (Heil, 1999, p. 26). AP is also often equated with the idea that philosophy is a continuation of science (Heil, 1999, p. 26). I do not claim to unite the social and the natural sciences within the covers of this book, but there is still value in the attempt to use logic and philosophical analysis to dispel some of the confusion surrounding liberty and technology.

Our everyday language tends to be ambiguous, which leads us into trouble whenever we use such words as *freedom* or *power* without defining them. I partly draw on the use of logic and formal methods in order to achieve some clarity of argument. The main tool of analytical philosophers is philosophical analysis, which involves the analysis of concepts and propositions instead of relying solely on language. This, it is hoped, provides a way towards laying "bare the logical form of reality" (Heil, 1999, p. 26).

Analysis of concepts lets us unpack questions, enabling us to get to their core and answering them. It involves the decomposition of something into its constituents (Stroll, 2000). Once the constituents are known, it becomes possible to construct logical arguments that show how concepts are related to one another. The advantage of such decomposition is that it enables us to tackle "problems one at a time, instead of having to invent at one stroke a block theory or a whole universe" (Russell, 1945, p. 834). Furthermore, this decomposition lets us identify any implicit reliance on dubious, or unclear, concepts (Heil, 1999).

A problem with much academic, and popular, debate is that words and conceptions are used without a common understanding of what the concepts really entail. The lack of such an understanding precludes both real understanding and proper debate (Stroll, 2000). Devoting a whole book to the linkage between Big Data and liberty provides the necessary space to properly define and develop these concepts—space that is rarely found in neither academic nor news articles. In this respect, the book is similar to the recent *Power & Technology*, in which Sattarov (2019) laments the confused usage of power in debates about the effects of technology. The same goes for liberty, as many have written of the impacts of Big data or AI on *freedom*, but rarely are the concepts properly defined and understood. My goal is to foster a common understanding of the concepts discussed, in order to promote and enable what Arne Næss (2016) calls *sober* debate. In such debates superficial agreements or "skin-deep disagreements" are done away with through definitions and proper explanations of the concepts we employ (Næss, 2016).

As my main methods employed are those of conceptual analysis and thought experiments, this book will disappoint some of those with a heavy empirical or technical bent. My aim is not to enter into highly technical debates of the technologies in question. I work from common definitions of certain technologies toward more general conceptions in which I focus on what I perceive to be their essential characteristics. My goal is to provide analyses that will be relevant both for the technologies as they exist today, *and* as they might exist for some time to come. I will, accordingly, only get into details of computer science and engineering when strictly necessary.

As I analyze key *concepts*, and formulate *propositions* and arguments in logical form, I perform what is called conceptual analysis (Foley, 1999). My conceptual analysis consists in first unpacking and defining the core concepts, and then constructing arguments along the lines of *if liberty is C, and technology D is characterized by E, technology D negatively affects liberty.* Such statements are *analytical* statements, the truth of which is determined by logic, and not by correspondence with reality (Godfrey-Smith, 2003). I will, however, also make *synthetic* statements, with claims that the concepts and arguments I employ *are* related to the empirical world (Godfrey-Smith, 2003). My main focus, however, is on the analytical parts, and while I consider my efforts to be relevant to empirical reality, I believe my main contribution is to provide the ground for further empirical research in order to determine and expound the synthetical aspects of the topics I analyze. In the Chapters 4, 5, and 6, I propose a set of propositions, which I subsequently analyze in order to determine what consequences they entail.

While the empirical reality has a great many benefits, there are times when counterfactuals are even better at making the fundamental details of concepts clear. One kind of counterfactual is the thought experiment (Dowding, 2015). In a thought experiment we construct fictional extreme cases in order to show how far certain principles apply, or do not apply. Dowding (2015) states that such experiments provide data for the nonempirical conceptual analysis. Stephen Van Evera (1997), however, notes that this data cannot replace empirical facts in the service of hypothesis testing. I agree with this statement, but my purpose is to construct, rather than test, theories of how technology may affect liberty. In this case Van Evera (1997) agrees counterfactuals can be useful.

When defining concepts, I aim to do so in a nonnormative manner. *If* liberty is such-and-such, it may be negatively affected by a certain technology. In my arguments, however, it is not necessary to assume that liberty is *good*. Enemies and lovers of liberty alike should, ideally, be forced by the concepts and logical arguments into the same conclusions, such as *if liberty is X, liberty is negatively affected by Y*. People of various normative attitudes can then continue the argument and say either that liberty is unimportant or that they prefer a completely different conception of liberty. As such, the analytical content of much of this book is descriptive and not normative (Godfrey-Smith, 2003).

However, even if I argue in such terms as *if X is Y, Z is W*, there are always, either officially or unofficially, normative elements contained in the theories and conclusions drawn (Godfrey-Smith, 2003). I also note that many *do* consider liberty to be both good and an essential part of the *good society* (Coeckelbergh, 2018; Griffy-Brown, Earp, & Rosas, 2018). For Leo Strauss (1988), for example, political philosophy *is* the quest for the good—and the good society.

Furthermore, when we move on from the mainly analytical parts of the book to the forays into its synthetic parts, normative aspects are unavoidable. I acknowledge this, but still state as clearly as I can that my intention is to show the main ways in which liberty may be negatively affected by Big Data, while not answering, for example, the question of whether utility or liberty is the greater good, when these inevitably collide. While there are normative elements in any political theoretical analysis, I aim to provide a descriptive and analytical basis for subsequent normative evaluations of these issues, well aware of the fact that I am not involved in a *purely* analytical and descriptive enterprise.

Who is this book for?

The preceding considerations reveal a desire to introduce and ground the current undertaking aimed primarily at those that come from a different background than my own. But who is this book really for? My goal is that political philosophers will learn something about technology, that technologists and those from other social sciences will learn something about liberty, and that interested readers in general might learn something about both these things *and* how they are connected.

This book is intended to promote interdisciplinary discussions of the consequences of technology. If people are to have reasoned debates about the effects of technology, they need to understand what the technologies entail, and what consequences they have. By explaining the technologies and defining general concepts describing their essential character, I hope to contribute to broadening the political debates about these issues. Similarly, by explaining the concepts of liberty, I simultaneously contribute to making these terms more available to the experts of technology and communication studies. If I manage to achieve at least *some* of these objectives, I will consider my project a success, as these debates are central in today's societies, and they will only become more important as time passes.

Analytical philosophy is at times both highly technical and inaccessible to the general public. This has led to criticism of the professionalization of philosophy and a "call for a return to a pluralistic, community-oriented style of philosophizing" (Heil, 1999, p. 26). I support this call, and aim to employ the tools of logical and conceptual analysis without making the book inaccessible. In doing so, I break with Bertrand Russell's advice for early career researchers. He suggests that young professors should write their first work "in a jargon only to be understood by the erudite few" in order to prove that they *can*; this hopefully being beyond dispute, they can later on write *well*—in ways all will understand (Russell, 2009, p. 37).

The interdisciplinary nature of this book, and the broad audience I aim for, will disappoint some specialists, abut hopefully help most others. One set of scientific ideals is *communism* (also referred to as *communality* or *communalism), universalism, disinterestedness,* and *organized skepticism* (Merton, 1973). Communism and universalism, interpreted broadly, imply that research should be widely accessible, and have social value. Ragnvalg Kalleberg (2007, p. 155) argues that several of today's most important societal challenges require that multiple sciences must contribute to the public debate, and that "mutual popularization" must take place between disciplines.

Facing the challenges connected to technology requires such mutual popularization. Let me at the outset state clearly that I do not claim to be an expert in all the fields I touch upon in this book. This entails a need to rely on good scientific "translations" of the research done in a range of disciplines in which I am not an expert. However, I also take part in this communal endeavor as I translate research from my own and other disciplines. Scientific humility is essential for both "learning and cognitive" development in the processes of doing such interdisciplinary work (Kalleberg, 2007, p. 155).

Finally, I hope that the book will promote *public* debate on the topics discussed. This is why I have tried to write in an accessible and nontechnical manner, keeping the complexity of the propositions and structure of the logical underpinnings to a minimum. Kalleberg (2007) points to Merton's linking of the ethos of science and the ethos of democracy. In addition to the similarities between them, one ethical aspect of science entails the willingness to make science *accessible* and *useful* for public discourse. Determining the possible political effects of technologies is a daunting task for anyone not familiar with technology, so how can this challenge be faced?

(a) Requiring the policymakers to be experts in the disciplines involved;

(b) Letting experts from *one/some* of the disciplines form policy;

(c) Doing our best to make the basics of the various fields accessible and relying on the democratic debate between experts and nonexperts to be the basis of policy.

The first option seems unfeasible, and while the second option might be efficient it is not necessarily particularly democratic (Sætra, 2020). A technocratic solution in which one or some of the disciplines are given authority will be met with opposition both from adherents of democracy and popular control of policy, and of course from the experts in the field that are *not* given authority (Danaher, 2016). Furthermore, the power of technology, and how influential it is in shaping our societies, necessitates both understanding and engagement by social scientists (Crawford et al., 2019). The challenges posed by new technologies are, as David J. Gunkel (2019) states, simply "too important and influential for us to leave decision-making to a few experts" and "we also need an informed and knowledgeable public."

The third option above thus appears to be the most attractive one. It is important that we make the research performed in the various relevant fields accessible both to experts from other fields and the public in general.

The structure of the book

The first task at hand is to develop and explain the key concepts I rely on, namely Big Data and liberty. Chapter 2 consists of a brief discussion of these concepts, and also the most important related concepts. Liberty, for example, is intimately related to the concept of power, and being able to distinguish such concepts from each other will provide the means to analyze how liberty is affected by technology. Similarly, Big Data can refer to a number of different concepts, most of which are tightly integrated with other concepts such as AI, machine learning, and algorithms.

Armed with an understanding of the key concepts, I turn to the first threat in Chapter 3. The main question asked is whether privacy is a requirement for liberty. Proponents of certain forms of liberalism in particular have struggled to explain just *why* privacy is important, and why surveillance might be inimical to liberty. Opponents of liberalism are perhaps even more keen to show why liberal theory provides insufficient means to grasp the threats posed by Big Data, but I argue that surveillance constitutes a form of interference that is—or should be—problematic for liberals and nonliberals alike.

Surveillance leads to the creation of data—and how this data is used to influence our actions is the topic of Chapter 4. When "nudging" is combined with Big Data, does it become manipulative in ways inimical to liberty or even coercive? I argue that it does, and the threat of Big Data-based nudging is the second threat. This highlights another area in which parts of the liberal tradition have arguably provided insufficient means to object. Influence and manipulation, after all, are by some argued not to constitute the interference that infringes upon liberty. I object once more: psychological force is both real and constitutive of a threat to liberty when it becomes effective enough.

The last threat concerns the algorithmic curation of information, and this is analyzed in Chapter 5. Data is used to control the information we receive in a variety of ways, and Big Data is at the core of modern practices of information curation. But what is the relationship between Big Data-based curation of information and liberty? The threats posed by algorithmic curation is somewhat different from the other two, and the focus of this chapter is shifted to issues of individuality, unimpeded access to the world around us, and the *tyranny* I argue to exist in the filter bubbles and echo chambers that can result from the application of algorithms.

The three threats are seemingly different, and for some not even that disconcerting when seen in isolation. In Chapter 6 I show how we should take an extra step and consider how these three threats *in concert* constitute a broad and relentless attack on liberty. They are in fact tightly related, and understanding these relationships allows us to glean the potential solutions available to us.

How should we face the threats described? In Chapter 7 we head toward potential means of resistance, and I argue that privacy is in fact the key factor in preventing the most unfortunate effects of Big Data. The nature of privacy is here examined, and I provide a proposal for a new and restated concept of liberty—built on traditional liberal theory, but updated and developed in order to account for the threats discussed in this book. Finally, I argue that political action—perhaps quite unpopular action— is required to prevent and reverse the effects of Big Data on individual liberty and our societies.

References

Barley, S. R. (2020). *Work and technological change.* USA: Oxford University Press.

Baruh, L., & Popescu, M. (2017). Big data analytics and the limits of privacy self-management. *New Media & Society, 19*(4), 579–596.

Beaney, M. (2013). *The Oxford handbook of the history of analytic philosophy.* Oxford University Press.

Bello-Orgaz, G., Jung, J. J., & Camacho, D. (2016). Social big data: Recent achievements and new challenges. *Information Fusion, 28,* 45–59.

Binns, R. (2018). Fairness in machine learning: Lessons from political philosophy. In *Paper presented at the conference on fairness, accountability and transparency.*

Buolamwini, J., & Gebru, T. (2018). *Gender shades: Intersectional accuracy disparities in commercial gender classification.* Paper presented at the conference on fairness, accountability and transparency.

Chen, H., Chiang, R. H., & Storey, V. C. (2012). Business intelligence and analytics: From big data to big impact. *MIS Quarterly,* 1165–1188. https://doi.org/10.2307/41703503

Coeckelbergh, M. (2018). Technology and the good society: A polemical essay on social ontology, political principles, and responsibility for technology. *Technology in Society, 52,* 4–9.

Cohen, J. E. (2012). What privacy is for. *Harvard Law Review, 126,* 1904.

Collingridge, D. (1980). *The social control of technology.* London: Frances Pinter.

Crawford, K., Dobbe, R., Dryer, T., Fried, G., Green, B., Kaziunas, E., … Sánchez, A. N. (2019). *AI now 2019 report.* New York, NY: AI Now Institute.

Danaher, J. (2016). The threat of algocracy: Reality, resistance and accommodation. *Philosophy & Technology, 29*(3), 245–268.

Dormehl, L. (November 29, 2019). *Surveillance on steroids: How A.I. is making Big Brother bigger and brainier.* Retrieved from https://www.digitaltrends.com/cool-tech/ai-taking-facial-recognition-next-level/.

Dowding, K. (2015). *The philosophy and methods of political science.* Macmillan International Higher Education.

Foley, R. (1999). Analysis. In R. Audi (Ed.), *The Cambridge dictionary of philosophy.* Cambridge: Cambridge University Press.

Godfrey-Smith, P. (2003). *Theory and reality: An introduction to the philosophy of science.* Chicago: University of Chicago Press.

Greenfield, P. (March 26, 2018). *The Cambridge Analytica files: The story so far.* The Guardian. Retrieved from https://www.theguardian.com/news/2018/mar/26/the-cambridge-analytica-files-the-story-so-far.

Griffy-Brown, C., Earp, B. D., & Rosas, O. (2018). Technology and the good society. *Technology in Society, 52,* 1–3.

Gunkel, D. J. (2019). *How to survive a robot invasion: Rights, responsibility, and AI.* Routledge.

Heil, J. (1999). Analytic philosophy. In R. Audi (Ed.), *The Cambridge dictionary of philosophy.* Cambridge: Cambridge University Press.

Heilbroner, R. (1994). Technological determinism revisited. In L. Marx, & M. R. Smith (Eds.), *Does technology drive history* (pp. 67−78). Cambridge: MIT Press.

Kalleberg, R. (2007). A reconstruction of the ethos of science. *Journal of Classical Sociology, 7*(2), 137−160.

Köchling, A., & Wehner, M. C. (2020). Discriminated by an algorithm: A systematic review of discrimination and fairness by algorithmic decision-making in the context of HR recruitment and HR development. *Business Research,* 1−54.

Latour, B. (1999). *Pandora's hope: Essays on the reality of science studies.* Cambridge: Harvard University Press.

Marx, L., & Smith, M. R. (1994). *Does technology drive history? The dilemma of technological determinism.* Cambridge: MIT Press.

McKew, M. K. (February 16, 2018). *Did Russia affect the 2016 election? It's now undeniable.* Wired. Retrieved from https://www.wired.com/story/did-russia-affect-the-2016-election-its-now-undeniable/.

Merton, R. K. (1973). The normative structure of science. In N. W. Storer (Ed.), *The sociology of science: Theoretical and empirical investigations.* Chicago: University of Chicago Press.

Næss, A. (2016). *En del elementære logiske emner.* Oslo: Universitetsforlaget.

Noble, S. U. (2018). *Algorithms of oppression: How search engines reinforce racism.* New York: New York University Press.

Olson, P. (2019, December 3). *With brits used to surveillance, more companies try tracking faces.* Wall Street Journal. Retrieved from https://www.wsj.com/amp/articles/with-brits-used-to-surveillance-more-companies-try-tracking-faces-11575369002.

Russell, B. (1945). *History of western philosophy.* Simon & Schuster.

Russell, B. (2009). How I write. In R. E. Egner, & L. E. Denonn (Eds.), *The basic writings of Bertrand Russell.* London: Routledge.

Sætra, H. S. (2020). A shallow defence of a technocracy of artificial intelligence: Examining the political harms of algorithmic governance in the domain of government. *Technology in Society,* 101283.

Sætra, H. S. (2021a). AI in context and the sustainable development goals: Factoring in the unsustainability of the sociotechnical system. *Sustainability, 13*(4), 1738. https://doi.org/10.3390/su13041738

Sætra, H. S. (2021b). Confounding complexity of machine action: A Hobbesian account of machine responsibility. *International Journal of Technoethics, 12*(1).

Sattarov, F. (2019). *Power and technology: A philosophical and ethical analysis.* Rowman & Littlefield.

Schwab, K. (2017). *The fourth industrial revolution: Currency.*

Strauss, L. (1988). *Persecution and the art of writing.* Chicago: University of Chicago Press.

Stroll, A. (2000). *Twentieth-century analytic philosophy.* Columbia University Press.

Van Evera, S. (1997). *Guide to methods for students of political science.* Cornell University Press.

Vinuesa, R., Azizpour, H., Leite, I., Balaam, M., Dignum, V., Domisch, S., ... Nerini, F. F. (2020). The role of artificial intelligence in achieving the Sustainable Development Goals. *Nature Communications, 11*(1), 1−10.

Wyatt, S. (2008). Technological determinism is dead; long live technological determinism. *The Handbook of Science and Technology Studies, 3,* 165−180.

Yeung, K. (2017). 'Hypernudge': Big Data as a mode of regulation by design. *Information, Communication & Society, 20*(1), 118–136.

Youyou, W., Kosinski, M., & Stillwell, D. (2015). Computer-based personality judgments are more accurate than those made by humans. *Proceedings of the National Academy of Sciences, 112*(4), 1036–1040.

Zuboff, S. (2019). *The age of surveillance capitalism: The fight for a human future at the new frontier of power: Barack Obama's books of 2019.* New York: PublicAffairs.

2

Technologies and society

Introduction

A threat is posed *by* something, *to* something. In this book the source of the threat is what I label *Big Data*, and the threat applies to *liberty*. Before detailing the threats, it is necessary to define these two concepts. Both of the concepts are subject to a wide array of different definitions and usages, and if this step is skipped, there is little hope of reaching agreement on whether or not the threats are real and serious.

The main purpose is not to determine what the correct definitions are, but simply to make clear how the concepts are used in this book. With regard to Big Data, I'll first make clear how I use the term before detailing the various underlying concepts and technologies involved. With liberty I take a slightly different approach. Rather than making clear how I use the term, I establish a set of common understandings of the concepts, including negative liberty, positive liberty, and republican liberty. These are all considered valid and important conceptions of liberty, and my approach to liberty is thus pluralistic. By acknowledging them all I'll show how the different threats affect the various types of liberty in different ways.

If at the end of this chapter we are in agreement on what Big Data and liberty refer to *in this book*, the purpose of the chapter has been achieved. We may then proceed to examine whether or not we agree on my analysis of the linkages between the concepts, or you may argue that other definitions or conceptualizations would have been more useful or interesting. Either way, we will have done what we can to prepare the ground for proper debate (Næss, 2016).

Big Data as a logic and a system of technologies

In order to analyze technological change, one may (a) focus on the specifics of existing technologies (or technologies that have existed) or (b) analyze the fundamental principles of the

Big Data's Threat to Liberty. https://doi.org/10.1016/B978-0-12-823806-6.00001-6
Copyright © 2021 Elsevier Inc. All rights reserved.

technologies in question. I have chosen the latter approach. This means that my approach is analytical and conceptual, and not empirical. For example, let us assume that I desire to analyze how Facebook might lead to the creation of echo chambers—communities in which like-minded people congregate online. If I followed approach (a), I would have to examine the specific algorithms that Facebook employs, in order to show how they may, in practice, lead to the phenomena in question. This is the approach taken by, for example, Taina Bucher (2012), as she analyzes how Facebook's algorithms may create a threat of invisibility. While this is of great importance, the problem with such an approach to the analysis of the implications for, say, liberty, is that it might become obsolete the moment Facebook updates its algorithms. Indeed, this research may even become the *reason* why Facebook changes its algorithms. If so, it would obviously have great value, but it would not be as useful for analyzing the general implications of technological change over time, in a broader perspective. In addition, such an approach requires detailed knowledge of the intricacies of information technology and programming, as well as access to the actual code of the various algorithms. Gaining access on this level is difficult, as these algorithms are the core of the companies in question, and they are considered trade secrets. Furthermore, the algorithms are continually tweaked and refined.

I favor the analytical approach, because if we correctly identify the main mechanisms involved in Big Data and AI, we can analyze the implications relevant for the past, the present, and potentially the future. Research based on such a foundation will be relevant as long as the technologies are based on the same fundamental principles. My aim is to provide an analysis that will stay relevant for as long as the tendency persists to gather massive amounts of data and employ them in an ever-changing array of ways in order to provide both private profit and social benefits.

The first approach described, that of focusing on specifics, is of great importance, but I believe that this is the proper domain of the discipline of information technology and other disciplines involved in the technical aspects of the phenomena discussed. The technical level is not the domain of political theory, but political theorists *must* be involved at the general level, where the overall implications of the interplay between man and technology are debated.

Big Data

Our age is often referred to as the *era of big data* (Boyd & Crawford, 2012; Chen, Chiang, & Storey, 2012; Sivarajah, Kamal, Irani,

& Weerakkody, 2017). If that is indeed the defining aspect of our times, we must know what Big Data *is*, and what implications it has for individuals and our societies. New technologies rarely emerge from philosophy or the social sciences, so we might expect some delay between the arrival of new technologies and the analysis of the implications for man and society. An interesting point to note is that a lot of technologies have been analyzed *before* their actual arrival, through such arenas as the combination of science fiction and philosophy (Schneider, 2016). However, when it comes to real phenomena, such as Big Data, there has been a lack of fundamental analysis of the implications and nontechnical aspects of it in academic arenas (Gandomi & Haider, 2015). When a new technology becomes both popular and profitable, the experts on the technology have incentives to focus on other channels than the traditionally somewhat slow academic journals. Gandomi and Haider (2015, p. 137) describe how experts often "leapfrog to books and other electronic media for immediate and wide circulation of their work" As a result, the understanding of the political theoretical aspects of Big Data is underdeveloped. This book is an attempt to address this issue.

Doug Laney (2001) identified the emerging tendency to gather increasing amounts of data, at an increasing speed, structured in heterogeneous data sets. These aspects of data, that we now call *Big Data*, are often referred to as the *three Vs*: volume, velocity, and variety. Today, Laney's original three Vs have been supplemented by a fourth, *value*, and sometimes a fifth, *veracity*. Value refers to how we extract valuable information from the data sets, and veracity refers to how correct and accurate the data is, as the data sets are often incomplete in a variety of ways (Bello-Orgaz, Jung, & Camacho, 2016; Helbing, 2015). The *term* Big Data was coined in 2005 by Roger Mougalas, when he used it to describe "a wide range of large data sets almost impossible to manage and process using traditional data management tools" (Halevi & Moed, 2012).

Bello-Orgaz et al. (2016) emphasize that Big Data is something *new* and not simply more of what has been before. Classical methods of handling data cannot be used to handle the growth of data and the processing and analysis of it (Bello-Orgaz et al., 2016; Manyika et al., 2011). This definition is intentionally a moving definition, as the technologies used to analyze the data continually change, and the size of the data sets vary between different sectors (Manyika et al., 2011).

Baruh and Popescu (2017) describe how data is collected both by voluntary provision and through obscure or automatic mechanisms. Shoshana Zuboff (2019) provides a comprehensive analysis of the extent to which data (with a focus on personal data) is gathered in today's society. Of particular interest is her discussion of how companies like Google and Facebook deploy every available method in their search for more data to feed their prediction engines, which is the core of their businesses (Zuboff, 2019). While private companies are important, I also include government use of Big Data in my understanding of the concept. Also, I mainly focus on Big Data related to *personal information* and *social relations*, as these kinds of data most clearly have the ability to affect liberty. When Big Data is discussed as a threat to liberty, I refer to this limited conception of Big Data, and not, for example, Big Data about the weather, or space, unless such data is used to influence the actions of individuals.

Speaking of value, in its 2018 Christmas edition, The Economist (2018) ran a leader on *How to think about data in 2019*, citing the now well-known saying that *data is the new oil*. Data, it is argued, is of little use until it is processed, and the goal of processing data is usually to get to know individuals. As Big Data is the new oil, it must be drilled, mined, and refined (Helbing, 2015, pp. 75–76). Of the seven largest companies in the world, most of them are based on "tying data to human beings" (The Economist, 2018). *Big Tech* is another name for the main companies involved in Big Data, and GAFA is an acronym often used for the four largest companies: Google, Apple, Facebook, and Amazon (Foer, 2017). Big data has become a highly interdisciplinary phenomenon, and it is now employed by "[c]omputer scientists, physicists, economists, mathematicians, political scientists, bio-informaticists, sociologists, and other scholars" (Boyd & Crawford, 2012, p. 662).

Of the recent efforts to describe the phenomenon of Big Data, Zuboff's (2019) previously mentioned *Surveillance Capitalism* deserves special mention. In this book, she describes the growth of the large technological companies, and how they purportedly manifest a change in capitalism itself, as data (personal data) is now what defines the industry. Data is indeed the "new oil" and by collecting as much data as possible, they aim to capture what Zuboff (2019) calls *behavioural surplus*—information that lets the companies use the data they gather for other purposes than simply improving their services (which is behavioral feedback). While Zuboff's book is of great interest, Cohen (2019) gives more detail on the legal aspects of the issues in question in her review of the book. Franklin Foer (2017) provides an account of the monopolistic tendencies of large companies, and Jaron Lanier (2014) goes farther than most when he asks *"Who owns the future?,"* and drafts a radical new digital order that respects the individual's contribution to the Big Data economy.

I acknowledge that Big Data provides great benefits in many areas of our daily lives, and to society in general. However, I also argue that we incur some costs when we employ Big Data in the way we do. Big Data is used in all aspects of our modern existence, from social media, to all areas of business, science, and even government (Baruh & Popescu, 2017; Chen et al., 2012). Andrej Zwitter (2014, p. 5) discusses the ethical aspects of Big Data, and states that "Big Data might induce certain changes to traditional assumptions of ethics regarding individuality, free will, and power" The threat to individuality is discussed in Chapter 5, while I have previously discussed the two other topics in Sætra (2019a, 2019b).

To sum up, when I use the concept Big Data I refer to a compound concept related to the technologies I discuss in the following. I consider Big Data to be the most descriptive term, and throughout the book I refer to the compound concept here defined. Big Data contains personal and social data, *and* related information that can be used in combination with the aforementioned data. Big Data is massive in scope, and it both demands and enables new techniques of analysis. This means (a) that the keepers of data can go more in depth and reap more insight from the data and (b) that these processes become opaquer, as the algorithmic machine learning processes leading to certain outcomes in the analysis are beyond the understanding even of experts (Sætra, 2018). As such, the analysis of Big Data often involves sending data through a *black box* in order to get certain output. Furthermore, Big Data can be, and *is*, held both by governments and private entities. I mostly focus on how private companies gather and use Big Data, but governments are also increasingly active in the use of Big Data for surveillance and control purposes. There are, however, important differences in how countries approach the new possibilities provided by Big Data, and the attitudes of citizens of various nations toward such use of Big Data (LaBrie, Steinke, Li, & Cazier, 2018).

Artificial intelligence, algorithms, and machine learning

While I use the term Big Data, my concept is intimately connected to artificial intelligence, algorithms, and machine learning. I consider these concepts only in so far as they are connected to the gathering of, analysis of, and employment of Big Data as just defined.

Artificial intelligence (AI) is a term used to describe systems "capable of performing tasks commonly thought to require intelligence" (Brundage et al., 2018, p. 9). Another definition, less reliant on human standards of comparison, implies that intelligence involves the capability of adapting behavior in order to reach a certain goal in various environments (Bäck, Fogel, & Michalewicz, 2018).

AI can be used to perform relatively simple routine tasks in, for example, industrial settings, as tasks that require intelligence do not necessarily require *high* intelligence. However, AI is also used to solve complex tasks, and it increasingly often exceeds human capacities. An example is how Deep Blue beat Garry Kasparov in chess in 1997 (Campbell, Hoane Jr, & Hsu, 2002). In 2016, another milestone that attracted a lot of attention was passed when *AlphaGo*, a computer program developed by Google's DeepMind to play the ancient board game of Go, finally managed to beat a top human professional (Chouard, 2016). *DeepMind* is the self-proclaimed "world leader in artificial intelligence research and its application for positive impact"; playing Go and now also chess with AlphaZero are just two of many applications (Google, 2020b, 2020c). A more recent DeepMind engine is *AlphaStar*, which recently made the rank of *grandmaster* in the game of StarCraft II—the latest on the list of challenges where human beings have thus far exceeded the capacities of AI (Google, 2020a).

The quest for machine intelligence is not new. Alan Turing, in 1950, wrote *Computing Machinery and Intelligence*, which is also the origin of the famous Turing test, a method for assessing whether a machine can be said to be intelligent (Turing, 2009). Turing himself worked in the field of AI, and managed to use a computer to decode the German "Enigma" code during the Second World War (Copeland, 2014).

AI is now employed in all facets of computing, and we encounter it every day in areas such as "speech recognition, machine translation, spam filters, and search engines" (Brundage et al., 2018, p. 9). AI controls airplanes, drives cars, assists semi-autonomous weapon systems, makes financial decisions, trades on the stock exchange, etc. (Kurzweil, 2015). Furthermore, AI is combined with Big Data in order to get more information from existing sources of surveillance, such as security cameras (Dormehl, 2019). Other examples include how AI is used to decide who gets bail and who does not, who gets loans, etc. (Tashea, 2018). AI is now even used in making hiring decisions (Crawford et al., 2019). Sherry Turkle (2017) describes how AI is now increasingly social, and interacts with children, adults, and the elderly, and I have previously written about various aspects related to social AI and social robots in Sætra (2019a, 2020a, 2020b, 2020c, 2021b).

I emphasize that I do not consider issues of *general intelligence* and other more speculative aspects of AI, such as the possibility of superintelligence, ultraintelligence, the singularity, or "high-level machine intelligence" (HLMI) (Boström, 2014; Good, 1966; Kurzweil, 2015; Müller & Bostrom, 2014). Gary Marcus & Ernest Davis (2019) provide a good argument for not being too worried about such capabilities of AI, and I have previously examined topics such as AI and creativity, general intelligence, the idea of a singularity, etc., in Sætra (2018, 2019a). While not being emphasized in this book, DeepMind and others are pursuing general algorithms, and the new *MuZero* is an example of the more promising attempts in this direction (Schrittwieser et al., 2020).

While AI is of great importance, my main business with the term comes from how it is applied to Big Data. Firstly, AI can be employed as a means of pure analysis of historical data. I return to this application shortly when we turn to machine learning. Secondly, and of more interest here, is how AI is used to predict, anticipate, and guide behavior based on the analysis of Big Data. How AI is used in this manner is discussed in both Chapters 4 and 5. In order to understand this aspect of AI and Big Data, two other terms are required: *algorithms* and *machine learning*.

Whenever a machine translates input into output, we use an algorithm—a term discussed in more detail in Chapter 5. Algorithms are often likened to recipes, consisting of a set of steps to reach a set goal (Kroll et al., 2016; Rader & Gray, 2015). When Facebook decides what you see in your newsfeed, an algorithm makes these decisions based on *affinity, weight*, and *time decay* (Bucher, 2012). We see more Facebook posts from those we often interact with, we see more posts of the type that we like, and we see more new posts than old ones. Based on all the available posts (input), the algorithm produces our newsfeed (output)—the computer has made a choice, so to speak. It is important to keep in mind that algorithms cannot be considered neutral, even if they are automated computer processes (Binns, 2018; Crawford et al., 2019; Dwork & Mulligan, 2013; Kroll et al., 2016; Noble, 2018; Rader & Gray, 2015; Sætra, 2018; Smith, 2019). *Some* logic is employed in order to decide how the input is processed, and the programmers of algorithms thus have great power to shape the world we see. Facebook's newsfeed promulgates the "explicit and implicit values of their designers" (Dwork & Mulligan, 2013, p. 35), and the humans behind the platforms and algorithms in question are held to account throughout this book (Sætra, 2021a).

In Chapter 5 I discuss how *filter bubbles* may arise when algorithms are designed to provide a user with the content they prefer

(Pariser, 2011). This can occur if an algorithm is designed and trained to increase the prevalence of content similar to the content the user has previously consumed. There is an ongoing debate about the presence of, and severity of, filter bubbles. Dominic Spohr (2017) hypothesizes that phenomena such as echo chambers and selective exposure are more important contributors to ideological polarization than filter bubbles. He calls for more research—"from every possible angle"—on the relationship between polarization, algorithmic curation, and selective exposure, and I heed his call in Chapter 5 (Spohr, 2017, p. 157). Filter bubbles might arise through seemingly innocuous algorithmic learning. However, Ali et al. (2019) show that Facebook's delivery algorithms can exacerbate such tendencies with regard to politically diverse content by making it more expensive for advertisers to reach audiences that are believed not to agree with the advertisers. People thus get more ads for political actors than they already agree with.

Bucher (2018) provides an important discussion of the history and technical aspects of algorithms, as well as the issue of algorithmic power and politics. Tarleton Gillespie (2010) suggests that tech companies have become "the primary keepers of the cultural discussion"—an important development, considering the fact that these companies have profit as their motive, and are as yet faced with little regulation (Zuboff, 2019). Foer (2017) discusses much of the same in his book *World without Mind*.[1]

One way to define intelligence is to focus on the possibility to learn, and this is what *machine learning* is all about. The concept refers to systems that can "improve their performance on a given task over time through experience" (Brundage et al., 2018, p. 9). While it sounds fancy, machine learning is closely related to good old-fashioned pattern recognition (Bishop, 2006, p. 7). An example of machine learning could be an algorithm trained to recognize different categories of welfare recipients. Data, called training sets, are fed into the model, and in these data, we show the computer what sort of benefits each historical person in the data set has received (we have categorized them). As the algorithm goes through these cases, it (hopefully) learns to subsequently categorize new cases in a similar manner (Bishop, 2006).

Reinforcement learning involves optimizing some function by way of trial and error, instead of the supervised approach where we give the computer examples of what is desired (Sutton

---

[1]See Gillespie (2014) for a general discussion of the relevance of algorithms and *platforms*.

& Barto, 2018). Playing games, for example, can be learnt by reinforcement learning, where the computer knows the rules of the game in question, and plays against itself (a large number of times) in order to discover strong moves and the best plays (Bishop, 2006). Artificial neural networks imitate information processing in biological systems, with artificial neurons that imitate the way neurons in human brains signal to other neurons in complicated networks (Bishop, 2006). When these systems gain a certain complexity—become deeply layered—we often use the term deep learning (Sutton & Barto, 2018). Marcus and Davis (2019) question both neural networks' potential for further development toward general intelligence and their resemblance to the biological structure of the human brain. A meta-study seems to support such doubts about the potential for continued exponential growth of AI capabilities through deep learning (Hao, 2019).

A key advantage of Big Data is that we now get massive amounts of data that can be used as training sets, the result being that our computers become better at the tasks we give them. This is how all the technologies mentioned come together in the concept I call *Big Data*.

Liberty in its various guises

Without liberty life is a misery, Alexander Hamilton once argued (Alexander, Zenger, & Hamilton, 1736). If he was right, it is important that we do not lose sight of the implications that the technologies in question have on liberty. Our quest for ever more benefits from Big Data would indeed be of dubious value if people's lives were made miserable as a result.

Liberty is often assumed to be one aspect of the good society (Griffy-Brown, Earp, & Rosas, 2018). But what *is* liberty, and what type of liberty, if any, is threatened by Big Data? The term "liberty" has many meanings, and many claim, without ever defining the concept, that technology influences our liberty. However, if we do not have a clear idea of what kind of liberty we are discussing, we will obviously not be able to agree on what sort of effects Big Data has.

While I focus on some specific threats posed by Big Data, it is conceivable that even if these threats are real, *overall freedom*, as Ian Carter (1999) uses the term, may be increased by the use of new technologies. Overall freedom is a term for describing freedom *as such*, and it implies that liberty (as such) is merely an aggregation of *specific* freedoms. For Carter (1999), freedom consists of the ratio between available actions and available plus unavailable actions. This view of liberty does not strike me as a sufficient basis for determining what freedom a person has, and as a result, I will mainly work with richer concepts of liberty. This is not particularly restrictive, as even Isiah Berlin's (2002)[2] negative liberty is a broader concept than the one just mentioned. However, the idea of liberty as such a ratio leads to the possibility that overall freedom might be increased by the way we employ Big Data (in terms of the number of alternatives open to us), while liberty, understood more broadly, suffers.

Negative liberty

As a starting point, I employ Gerald C. Maccallum's (1967) general definition of liberty, as the freedom "of something (an agent or agents), from something, to do, not do, become, or not become something; it is a triadic relation" (p. 314). Carter (1999, p. 15) refers to this concept of freedom as "the absence of certain preventing conditions of agent's possible actions." In the general sense, then, we have an actor who either has or does not have freedom to perform some action. Freedom is thus determined by the presence or absence of *preventing* conditions, and what these may be is a crucial question in this book. While the brute force of an assailant who physically restrains you is certainly a preventing condition, imagining how surveillance and the manipulation of information in order to influence our actions constitute such preventing conditions requires us to define the concepts involved in more detail. Positive liberty is discussed in the next section, where the presence of *enabling* conditions become just as important as the absence of preventing conditions.

Liberty can thus be understood as the absence of external impediments, and this is the usual way to portray what Berlin (2002) labeled *negative* liberty. According to Berlin (2002, p. 169), negative liberty is "the degree to which no man or body of men interferes with my activity." If I can "act unobstructed" and I am "not being prevented by others from doing what I could otherwise do" I am free (Berlin, 2002, p. 169). Absolute negative liberty is impossible, as the actions of other people will by necessity lead to some contraction of my liberty. This does not, however, subtract from my liberty, but anything *beyond* the minimum of contraction makes me "coerced, or it may be, enslaved" (Berlin, 2002, p. 169). It is important to note that only the actions of other *people* are

[2]Berlin's *Two Concepts of Liberty* was first published in 1958.

considered obstacles to negative liberty (Berlin, 2002). Consequently, freedom from nature will not be considered. In this context, the actions of AI are considered the actions of human beings, in line with the argument I propose in Sætra (2021a).

We need not only consider morally culpable actions as impediments, but all actions where "other agents can be held *morally responsible*" Carter (1999, pp. 221, 236). If I run an online food store, I may not have a moral responsibility to make sure that the algorithms I employ dissuade you from purchasing unhealthy food. Thus, it would not be morally wrong for me to use these algorithms in ways leading you toward unhealthy habits, and there is no culpability to speak of. I may, however, still be held morally responsible for my choice of algorithms, without an accompanying adjudication of the moral justification for my choice Carter (1999, p. 235).

A *causal responsibility* approach to obstacles to freedom entails the attribution of responsibility for an effect to people who have in some way taken part in the causal chains that led to this effect. This leads to some potentially absurd situations, and I will not pursue this criterion. By this criterion, I might, for example, be assigned blame for the death of a pedestrian run over by a bus, if I approached in my car from the opposite direction and stopped in front of the pedestrian crossing, leading the pedestrian to start crossing and then being run over. If I merely followed the rules of traffic, and had no way of knowing what this would lead to, my actions are neither morally culpable nor am I morally responsible in any way.

Some, like Joseph Raz (1986), mainly consider the *intentional* actions of others. If I am once again a shopkeeper, and I had no intention of making you obese, all would be good. I find this view needlessly restrictive, as I may clearly have some responsibility for your obesity, despite this not being my intention. Neglect can also be a cause of moral attribution. I thus consider morally attributable acts to be of significance with regard to determining what sort of interference reduces liberty.

There are two things of particular importance to the analyses I provide: (a) Berlin's emphasis on the actions of others and (b) the fact that we might construe interference by others as encompassing more than the use of pure *physical* force. Regarding the first, Sunstein (2016), for example, argues that even nature nudges. If, as I argue, nudging is problematic with regard to liberty, there is an important distinction between the nudges of nature and the nudges of humans. As mentioned, I will only consider nudges that can be morally attributed to other people.

The other point relates to the possibility that the use of psychological force involves a contraction of negative liberty, something I discuss in more detail in Chapter 4, and also briefly in Chapter 5. Tom Goodwin (2012, p. 88) claims that proponents of negative liberty see it as an "abuse of words" to consider psychological pressure an obstacle to liberty, but this might be a premature conclusion. Faden and Beauchamp (1986, p. 355) use the term psychological manipulation, while Hopper and Hidalgo (2006) write about psychological coercion, and maintain that it "can be as effective as physical violence in exerting control over a person" (Hopper & Hidalgo, 2006, p. 186).

Positive liberty

Turning to *positive* liberty, this involves a desire "on the part of the individual to be his own master" (Berlin, 2002, p. 178). Carter (1999, p. 6) labels Berlin's positive liberty *self-mastery*, which involves the "the dominance of an 'authentic' self over a merely empirical self." We all have certain characteristics resulting from whatever circumstances made us who we are today. These are our empirical characteristics, and for someone concerned with self-mastery such characteristics might be of less interest than our authentic characteristics. Authentic characteristics are something akin to potential characteristics—things we might achieve (or others might promote in us through enlightenment or removal of preventing circumstances), and not merely something we have (Carter, 1999, p. 149). We thus speak of liberty as related also to the presence or absence of *enabling* conditions.

For me to have much positive liberty, my life would have to be the result of my own volition and actions, as far as possible. I am a *subject*—not an *object*. My liberty would be violated if some external influence prevented me from achieving what I otherwise would have achieved (Berlin, 2002, p. 178). This conception of liberty is of great interest with regard to the topics of Chapters 4 and 5. By nudging us, someone actively attempts to overrule, or change, the way we act. Arranging and re-arranging the information we receive about the world in different ways also influence our actions and decisions in a myriad of ways.

Being enslaved is intuitively an impediment for liberty in any form, but in the frame of positive liberty, the road to serfdom is quite short. It might be a slavery to nature, passions, or spiritual slavery, and the way out of slavery is emancipation, which leads to a "higher" understanding of the self. Berlin acknowledges that the aspects of liberty covered by positive liberty are important, but his article clearly warns against the dangers that follow from focusing on these aspects (Berlin, 2002). Berlin warns of two distinct ways in which positive liberty may lead us onto dangerous paths. Firstly, through self-abnegation in order to achieve independence and liberty, and secondly, through self-realization understood as attempting to gain liberty through "total self-identification with a specific principle or ideal" (Berlin, 2002, p. 181).

Berlin (2002) highlights the dangers of certain kinds of self-realization following a focus on positive liberty. Raz (1986), however, provides an example of a theory of positive liberty that does not adopt the aspects Berlin considers dangerous. Raz explicitly separates self-realization from autonomy, as "one can stumble into a life of self-realisation or be manipulated into it or reach it in some other way which is inconsistent with autonomy" (Raz, 1986, p. 375). Nowhere in Raz's writings on liberty are there hints of the "enlightened rationalism" or "higher self" that Berlin dreads. Autonomy does not necessarily involve being able to see things "as they really are" for example, and Raz's autonomy is clearly inimical to any ideology that legitimizes the use of coercion of individuals in order to help them achieve their "real" or higher, selves, or some sort of "pure rationality."

While Raz (1986) considers individuals as situated in society, and the ideal of moral individualism to be wrong, he does not slip down the slippery slope of Rousseau and others who portray freedom as the immersion in the whole. Raz recognizes the individual, while simultaneously recognizing that the individual requires society. He does not argue that liberty consists in ruling oneself through identifying with society as a whole. Nor does Raz's autonomy-based freedom seem to provide any impetus to self-abnegation. People's desires and interests are seen as genuinely valuable, and the idea that one might simply rid oneself of one's actual desires in order to become free is not the path toward autonomy-based freedom. I thus argue that it is possible to formulate a type of positive liberty which is not susceptible to the criticism of Berlin, and this will be developed further in Chapter 7.

Internal constraints are relevant for positive liberty, and such constraints may be things such as "fears, compulsions, obsessions, or limited aspirations" (Carter, 1999, p. 150). For us to call such phenomena constraints on action, we must have a value-based and non-empirical view of freedom, according to Carter (1999). Carter (1999) discourages such a view, while Raz (1986) is a proponent of a value-based theory of liberty. I do not consider efforts by others to free people from their natural internal constraints, but I do consider situations in which other people either cause or change such constraints, or attempt to exploit them.

I here follow Carter (1999, p. 221) and limit my analysis to constraints and actions "for which humans can be held morally responsible." These are cases in which threats make people, for example, experience fear, or in which people use nudging to exploit some irrational mechanism or internal constraint. While coercion involves interference with a person's options, manipulation "perverts the way that person reaches decisions, forms preferences or adopts goals" (Raz, 1986, pp. 377–378). Like coercion, such manipulation is regarded as a clear invasion of autonomy, and thus wrong, as freedom from both is a condition of autonomy. Unlike coercion, manipulation does not limit a person's physically available options. It consists in *perverting* the way a person makes decisions, in order to subject the will of the agent to that of the manipulator. This can be done through manipulating the decision-making process, preference formation, or goal adaption (Raz, 1986).

The perversion of our decision-making processes is emphasized in Chapter 4. This involves, for example, appeals to known personal biases or irrational proclivities, instead of appealing to a person's rational decision-making faculties. While we constantly make decisions under the influence of such irrational mechanisms, I focus on the morally attributable exploitation of such mechanisms when discussing manipulation as a threat to liberty.

Brainwashing is another interesting example of how psychological interference can be construed as inimical to liberty (Carter, 1999, p. 222). With regard to aspects of manipulation and the possibility of psychological coercion, such phenomena are obviously problematic for proponents of positive liberty (Raz, 1986). It is also interesting to consider such things as the nonneutrality of algorithms, and the power of their creators to influence us, as problematic. What is called affinity profiling, for example, involves grouping people based on assumed interests, and not solely their own personality traits (Wachter, 2020). According to Sandra Wachter (2020), such profiling may lead to unlawful discrimination. However, it also highlights the common good aspects of privacy, as the profiles of others are used to target similar individuals.

However, we might simultaneously entertain the idea that the use of manipulation can be *emancipatory*. This is related to the paternalism employed in the nudging of Thaler and Sunstein (2003, 2008). Imagine that I am enslaved by my passions and other irrational mechanisms. A well-directed nudge that lets me overcome such undesirable obstacles to good decisions might be construed as emancipatory.

For a fuller description of negative and positive liberty, see Berlin's *Two Concepts of Liberty* (2002) and Charles Taylor's (1985) critique of the concept of negative liberty in his article "What's Wrong with Negative Liberty." Gerald C. MacCallum (1967) and Quentin Skinner (2002) provide important critiques of the two concepts of liberty.

Liberty as nondomination, independence, and autonomy

Finally, I briefly introduce three additional types of liberty that will serve to broaden the analysis that follows, while also inspiring the restated concept of liberty that follows in Chapter 7.

The first additional take on liberty is the conception of liberty as *nondomination*. Domination, according to Philip Pettit (1997), consists in (a) the capacity to interfere, (b) arbitrarily, (c) in choices a person could otherwise make (Pettit, 1997). Skinner's (2002, 2008), conceptions of liberty are usually referred to as republican liberty (Carter, 1999). The main idea in the republican tradition is said to be an "escape from the arbitrary" which involves being subject to someone's "capricious will" and "idiosyncratic judgement" (Pettit, 1997, p. 5). According to Pettit, this kind of liberty lies *between* positive and negative liberty. As with negative liberty, it does not focus on self-mastery, and it requires the absence of domination by others. However, it goes further than negative liberty, since the mere absence of interference is not enough. The absence must be secured, which means that whenever someone has the opportunity to dominate us, this is a problem, even if they should choose not to (Pettit, 1997, p. 51).

With such an account of liberty, the questions I raise regarding the power of Big Data become highly relevant. Even if the people and companies controlling the algorithms that decide what world we will perceive do *not* abuse it, or do not abuse it *constantly*, it is still a problem that they have the power to do so. Furthermore, they have the power to do so arbitrarily. The very fact that such potential power exists violates Pettit's conditions of liberty, which require that people do not have the capacity to arbitrarily interfere with our actions. The techniques of nudging raise a similar concern, as these techniques, when combined with Big Data, also become powerful enough to violate the conditions just described. These issues are discussed in more detail in Chapters 4 and 5. While surveillance, discussed in Chapter 3, may seem more peripheral to this discussion of freedom as nondomination, I argue that privacy plays a pivotal role in denying actors the powers just discussed in relation to nudging and algorithmic power.

Closely related to republican liberty is liberty as independence, which is stated to be positioned between Berlin's negative liberty and Pettit's republican liberty (List & Valentini, 2016). While agreeing with Pettit that we need protection from potential interference, they disagree with proponents of republican liberty in the latter's emphasis on freedom from *arbitrary* domination. In order to better capture (a) all situations of liberty reduction in need of justification and (b) common language usage of the word freedom, they believe that also any nonarbitrary liberty reduction must be seen as a loss of liberty (List & Valentini, 2016). A prisoner justly imprisoned will thus be deprived of liberty according to liberty as independence, but not, List and Valentini (2016) argue, necessarily according to republican liberty. Liberty as independence is thus a nonmoralized conception of liberty (as Berlin's negative liberty).

Finally, there is the idea that liberty is intimately connected to autonomy. In Joseph Raz's (1986) *The Morality of Freedom*, we see that liberty can be understood as dependent on autonomy. This requires that a person is "part author" of their own life, and that we do not only care about what a person and their lives are right now, but also about "the way it became what it is" (Raz, 1986, pp. 204, 369). This is related to Berlin's division between an empirical and authentic self. The first entails whatever a person is, whereas the latter involves a consideration of what a person *could* be (Berlin, 2002). Conceptions of liberty that include considerations of the authentic self are usually considered to be varieties of positive liberty, and the reason for singling out Raz's theory is his development of the *conditions of autonomy*, which are "appropriate mental abilities, an adequate range of options, and independence" (Raz, 1986, p. 372). The first condition is not emphasized in this book, but the second and third certainly are. The second relates closely to negative liberty, but the third condition is highly relevant to the consideration of the value of privacy and Chapter 7, as independence might require a certain level of privacy. Independence in this sense, which must be distinguished from the independence of List and Valentini (2016), requires the freedom "from coercion and manipulation by others" (Raz, 1986, p. 373).

References

Alexander, J., Zenger, J. P., & Hamilton, A. (1736). *A brief narrative of the case and trial of John Peter Zenger, printer of the New-York weekly-journal.* New York: W. Dunlap.

Ali, M., Sapiezynski, P., Korolova, A., Mislove, A., & Rieke, A. (2019). *Ad delivery algorithms: The hidden arbiters of political messaging.* arXiv preprint arXiv: 1912.04255.

Bäck, T., Fogel, D. B., & Michalewicz, Z. (2018). *Evolutionary computation 1: Basic algorithms and operators.* CRC press.

Baruh, L., & Popescu, M. (2017). Big data analytics and the limits of privacy self-management. *New Media & Society, 19*(4), 579–596.

Bello-Orgaz, G., Jung, J. J., & Camacho, D. (2016). Social big data: Recent achievements and new challenges. *Information Fusion, 28,* 45–59.

Berlin, I. (2002). Two concepts of liberty. In H. Hardy (Ed.), *Liberty.* Oxford: Oxford University Press.

Binns, R. (2018). Fairness in machine learning: Lessons from political philosophy. In *Paper presented at the conference on fairness, accountability and transparency.*

Bishop, C. M. (2006). *Pattern recognition and machine learning.* springer.

Boström, N. (2014). *Superintelligence: Paths, dangers, strategies.* Oxford: Oxford University Press.

Boyd, D., & Crawford, K. (2012). Critical questions for big data: Provocations for a cultural, technological, and scholarly phenomenon. *Information, Communication & Society, 15*(5), 662–679. https://doi.org/10.1080/1369118X.2012.678878.

Brundage, M., Avin, S., Clark, J., Toner, H., Eckersley, P., Garfinkel, B., … Filar, B. (2018). *The malicious use of artificial intelligence: Forecasting, prevention, and mitigation.* arXiv preprint arXiv:1802.07228.

Bucher, T. (2012). Want to be on the top? Algorithmic power and the threat of invisibility on Facebook. *New Media & Society, 14*(7), 1164–1180.

Bucher, T. (2018). *If... then: Algorithmic power and politics.* Oxford University Press.

Campbell, M., Hoane, A. J., Jr., & Hsu, F.-h. (2002). Deep blue. *Artificial Intelligence, 134*(1–2), 57–83.

Carter, I. (1999). *A measure of freedom.* Oxford: Oxford University Press.

Chen, H., Chiang, R. H., & Storey, V. C. (2012). Business intelligence and analytics: From big data to big impact. *MIS Quarterly,* 1165–1188. https://doi.org/10.2307/41703503.

Chouard, T. (2016). The go files: AI computer wraps up 4–1 victory against human champion. *Nature News.* https://www.nature.com/news/the-go-files-ai-computer-wraps-up-4-1-victory-against-human-champion-1.19575. (Accessed 15 March 2016).

Cohen, J. E. (2019). Review of Zuboff's the age of surveillance capitalism. *Surveillance and Society, 17*(1/2), 240–245.

Copeland, B. J. (2014). *Turing: Pioneer of the information age.* Oxford University Press.

Crawford, K., Dobbe, R., Dryer, T., Fried, G., Green, B., Kaziunas, E., … Sánchez, A. N. (2019). *AI now 2019 report.* New York, NY: AI Now Institute.

Dormehl, L. (November 29, 2019). *Surveillance on steroids: How A.I. is making Big Brother bigger and brainier.* Retrieved from https://www.digitaltrends.com/cool-tech/ai-taking-facial-recognition-next-level/.

Dwork, C., & Mulligan, D. K. (2013). It's not privacy, and it's not fair. *Stanford Law Review. Online, 66,* 35.

Faden, R. R., & Beauchamp, T. L. (1986). *A history and theory of informed consent.* Oxford: Oxford University Press.

Foer, F. (2017). *World without mind.* Random House.

Gandomi, A., & Haider, M. (2015). Beyond the hype: Big data concepts, methods, and analytics. *International Journal of Information Management, 35*(2), 137–144.

Gillespie, T. (2010). The politics of 'platforms'. *New Media & Society, 12*(3), 347–364.

Gillespie, T. (2014). The relevance of algorithms. In T. Gillespie, P. Boczkowski, & K. Foot (Eds.), *Media technologies: Essays on communication, materiality, and society*. Cambridge: MIT Press.

Good, I. J. (1966). Speculations concerning the first ultraintelligent machine. In *Advances in computers* (Vol. 6, pp. 31–88). Elsevier.

Goodwin, T. (2012). Why we should reject 'nudge'. *Politics, 32*(2), 85–92.

Google. (2020a). *AlphaStar: Mastering the real-time strategy game StarCraft II*. Retrieved from https://deepmind.com/blog/article/alphastar-mastering-real-time-strategy-game-starcraft-ii.

Google. (2020b). *AlphaZero: Shedding new light on the grand games of chess, shogi and go*. Retrieved from https://deepmind.com/blog/article/alphazero-shedding-new-light-grand-games-chess-shogi-and-go.

Google. (2020c). *Solve intelligence. Use it to make the world a better place*. Retrieved from https://deepmind.com/about/.

Griffy-Brown, C., Earp, B. D., & Rosas, O. (2018). Technology and the good society. *Technology in Society, 52*, 1–3.

Halevi, G., & Moed, H. (2012). The evolution of big data as a research and scientific topic: Overview of the literature. *Research Trends, 30*(1), 3–6.

Hao, K. (January 25, 2019). *We analyzed 16,625 papers to figure out where AI is headed next*. MIT Technology Review. Retrieved from https://www.technologyreview.com/2019/01/25/1436/we-analyzed-16625-papers-to-figure-out-where-ai-is-headed-next/.

Helbing, D. (2015). *Thinking ahead-essays on big data, digital revolution, and participatory market society* (Vol. 10). Springer.

Hopper, E., & Hidalgo, J. (2006). Invisible chains: Psychological coercion of human trafficking victims. *Intercultural Human Rights Law Review, 1*, 185.

Kroll, J. A., Barocas, S., Felten, E. W., Reidenberg, J. R., Robinson, D. G., & Yu, H. (2016). Accountable algorithms. *University of Pennsylvania Law Review, 165*, 633.

Kurzweil, R. (2015). Superintelligence and singularity. In S. Schneider (Ed.), *Science fiction and philosophy: From time travel to superintelligence* (pp. 146–170). Chichester: Wiley-Blackwell.

LaBrie, R. C., Steinke, G. H., Li, X., & Cazier, J. A. (2018). Big data analytics sentiment: US-China reaction to data collection by business and government. *Technological Forecasting and Social Change, 130*, 45–55.

Laney, D. (2001). 3D data management: Controlling data volume, velocity and variety. *META Group Research Note, 6*(70), 1.

Lanier, J. (2014). *Who owns the future?* Simon and Schuster.

List, C., & Valentini, L. (2016). Freedom as independence. *Ethics, 126*(4), 1043–1074. https://doi.org/10.1086/686006.

MacCallum, G. C. (1967). Negative and positive freedom. *The Philosophical Review, 76*(3), 312–334.

Manyika, J., Chui, M., Brown, B., Bughin, J., Dobbs, R., Roxburgh, C., & Hung Byers, A. (2011). *Big data: The next frontier for innovation, competition, and productivity*. McKinsey Global Institute.

Marcus, G., & Davis, E. (2019). *Rebooting AI: Building artificial intelligence we can trust*. Pantheon.

Müller, V. C., & Bostrom, N. (2014). Future progress in artificial intelligence: A poll among experts. *AI Matters, 1*(1), 9–11.

Næss, A. (2016). *En del elementære logiske emner*. Oslo: Universitetsforlaget.

Noble, S. U. (2018). *Algorithms of oppression: How search engines reinforce racism*. New York: New York University Press.

Pariser, E. (2011). *The filter bubble: What the internet is hiding from you.* Penguin UK.

Pettit, P. (1997). *Republicanism: A theory of freedom and government.* Clarendon Press.

Rader, E., & Gray, R. (2015). *Understanding user beliefs about algorithmic curation in the Facebook news feed.* Paper presented at the proceedings of the 33rd annual ACM conference on human factors in computing systems.

Raz, J. (1986). *The morality of freedom.* Oxford: Clarendon Press.

Sætra, H. S. (2018). Science as a vocation in the era of big data: The philosophy of science behind big data and humanity's continued part in science. *Integrative Psychological and Behavioral Science, 52*(4), 508–522.

Sætra, H. S. (2019a). The ghost in the machine. *Human Arenas, 2*(1), 60–78.

Sætra, H. S. (2019b). Man and his fellow machines: An exploration of the elusive boundary between man and other beings. In F. Orban, & E. Strand Larsen (Eds.), *Discussing borders, escaping traps: Transdisciplinary and transspatial approaches.* Waxman: Münster.

Sætra, H. S. (2020a). First, they came for the old and demented. *Human Arenas,* 1–19. https://doi.org/10.1007/s42087-020-00125-7.

Sætra, H. S. (2020b). The foundations of a policy for the use of social robots in care. *Technology in Society, 63,* 101383.

Sætra, H. S. (2020c). The parasitic nature of social AI: Sharing minds with the mindless. *Integrative Psychological and Behavioral Science,* 1–19.

Sætra, H. S. (2021a). Confounding complexity of machine action: A Hobbesian account of machine responsibility. *International Journal of Technoethics, 12*(1).

Sætra, H. S. (2021b). Social robot deception and the culture of trust. *Paladyn: Journal of Behavioural Robotics, 12*(1). https://doi.org/10.1515/pjbr-2021-0021.

Schneider, S. (2016). *Science fiction and philosophy: From time travel to superintelligence.* John Wiley & Sons.

Schrittwieser, J., Antonoglou, I., Hubert, T., Simonyan, K., Sifre, L., Schmitt, S., ... Graepel, T. (2020). Mastering atari, go, chess and shogi by planning with a learned model. *Nature, 588,* 604–609. https://doi.org/10.1038/s41586-020-03051-4.

Sivarajah, U., Kamal, M. M., Irani, Z., & Weerakkody, V. (2017). Critical analysis of Big Data challenges and analytical methods. *Journal of Business Research, 70,* 263–286.

Skinner, Q. (2002). *A third concept of liberty.* Paper presented at the proceedings of the British academy.

Skinner, Q. (2008). *Hobbes and republican liberty.* Cambridge: Cambridge University Press.

Smith, R. E. (2019). *Rage inside the machine: The prejudice of algorithms, and how to stop the internet making bigots of us all.* Bloomsbury Academic.

Spohr, D. (2017). Fake news and ideological polarization: Filter bubbles and selective exposure on social media. *Business Information Review, 34*(3), 150–160.

Sunstein, C. R. (2016). *The ethics of influence: Government in the age of behavioral science.* Cambridge: Cambridge University Press.

Sutton, R. S., & Barto, A. G. (2018). *Reinforcement learning: An introduction.* MIT press.

Tashea, J. (February 2, 2018). *Courts are using AI to sentence criminals. That must stop now.* Wired. Retrieved from https://www.wired.com/2017/04/courts-using-ai-sentence-criminals-must-stop-now/.

Thaler, R. H., & Sunstein, C. R. (2003). Libertarian paternalism. *American Economic Review, 93*(2), 175–179.

Thaler, R. H., & Sunstein, C. R. (2008). *Nudge: Improving decisions about health, wealth, and happiness.* NewYork: Yale University Press.

Taylor, C. (1985). *Philosophy and the human sciences: Philosophical papers 2.* Cambridge: Cambridge University Press.

The Economist. (December 22, 2018). *How to think about data in 2019.* The Economist. Retrieved from https://www.economist.com/leaders/2018/12/22/how-to-think-about-data-in-2019.

Turing, A. M. (2009). Computing machinery and intelligence. In *Parsing the turing test* (pp. 23–65). Springer.

Turkle, S. (2017). *Alone together: Why we expect more from technology and less from each other.* UK: Hachette.

Wachter, S. (2020). Affinity profiling and discrimination by association in online behavioural advertising. *Berkeley Technology Law Journal, 35*(2).

Zuboff, S. (2019). *The age of surveillance capitalism: The fight for a human future at the frontier of power: Barack Obama's books of 2019.* New York: PublicAffairs.

Zwitter, A. (2014). Big data ethics. *Big Data & Society, 1*(2). https://doi.org/10.1177/2F2053951714559253.

3

Liberty under surveillance

Introduction

In the *era of Big Data* someone is always watching us. The websites we visit, the products we like, who we find attractive, and how we move around—all this is observed, stored, and finally analyzed. Big Data is a kind of omniscient and ubiquitous presence, and I here examine whether or not Big Data threatens freedom, both in the negative and in the positive sense (Berlin, 2002). I argue that privacy is threatened by how we employ Big Data, and that the various forms of surveillance that ensue constitute a threat to liberty in its various forms.

I arrive at three propositions in the course of this chapter, and they will serve as premises for the argument that Big Data-based surveillance is inimical to liberty. These propositions are (a) Big Data threatens privacy and enables surveillance, (b) the lack of alternatives to lifestyles that involves feeding into Big Data leads to something akin to forced participation in the surveillance of Big Brother, and (c) surveillance and the lack of privacy are a threat to freedom, because (i) the information gathered can be abused and (ii) people have a right not to be observed (even if the surveillance is completely benign). Using these propositions as premises, I show that liberty, in both the negative and the positive sense, is threatened. It is a two-pronged threat to liberty since (a) we experience a lack of alternatives to taking part in the collection of Big Data, and (b) taking part involves a loss of liberty by placing us under surveillance. The positive conception of liberty provides the strongest argument against how we currently employ Big Data, but the negative conception can also provide a sufficiently strong argument in favor of limiting Big Data-based surveillance.

Big Data surveillance

The gathering of Big Data

We leave countless traces of our daily activities in the rapidly growing collections of databases. A multitude of companies and institutions gather the different snippets of data, and the whole

Big Data's Threat to Liberty. https://doi.org/10.1016/B978-0-12-823806-6.00005-3
Copyright © 2021 Elsevier Inc. All rights reserved.

35

structure that has been set up to gather data has been called the "surveillant assemblage" (Haggerty & Ericson, 2000). This term refers to the combined effects of private and public efforts, markets, and institutions that, together, form an assemblage of surveillance. Much of the information is gathered by private companies, but government "is an important secondary beneficiary" (Cohen, 2012, p. 1916; Solove, 2004).

I here work on the assumption that we can meaningfully consider Big Data a phenomenon that constitutes a unitary threat to liberty. Big Brother is a metaphor often used to illustrate the nature of modern society. Unlike Big Brother in Orwell's (1949) fictional Oceania, Big Data is not believed or supposed to be one single actor. Instead of one Big Brother, we might have many *small* (and some quite large) brothers, and this conglomeration of large and small, public and private, actors that form the basis of Big Data-based surveillance has led some to argue that the metaphor is ill-fitted for the purpose of explaining modern surveillance (Lyon, 1994; Solove, 2004).

However, unless one considers the way in which these little brothers act in concert, almost in unison, the overall effects of modern surveillance remain obscure (Sætra, 2019; Solove, 2004). Zuboff (2015, p. 81; 2019) uses the term big *other* to describe the beast of surveillance, and by this she refers to "a ubiquitous networked institutional regime that records, modifies, and commodifies everyday experience [...]."

As such, I argue that we can in certain contexts meaningfully refer to something akin to *a* Big Brother, and by this refer to the surveillant *assemblage* of actors. This assumption is not uncontroversial, and I fully accept that there is not *one* large company, organization, or government that collects or organizes the collection of data. However, I agree with the view of Haggerty and Ericson (2000, p. 605) that we are seeing a "convergence of once discrete surveillance systems"—not a convergence in the sense of corporate consolidation, but in the sense that the countless separate streams of information about individuals is subsequently combined, shared, traded, and reassembled to provide a comprehensive and full description of individuals and their actions. Big Data consists of a large number of surveillance sources that function as an *assemblage*—the unity of this Big Brother is not formal, but *functional* (Deleuze & Guattari, 1988).

From Big Data to Big Brother

In Alan Westin's (1967) *Privacy and Freedom*, the challenges to privacy from new technologies and public concern about the dangers of *Big Brother* were discussed. More than 50 years later, the book is arguably more relevant than ever before. Big Data has changed things, since Big Brother has never had a better chance to work his magic while both keeping an eye on what is happening and preventing what he does not like. Overt surveillance may have given way to covert surveillance—a form that feels quite cosy, and that is less dependent on authority and police enforcement and more on the apparently voluntary provision of data. Unfreedom can in fact be quite comfortable (Marcuse, 2013) We agree to terms most of us do not even read (Böhme & Köpsell, 2010). The data collected are today's gold, and we happily give it away to gain access to various services. We connect with people and companies, get news and laughs, and in living our lives this way, we provide all the information required to *really* know us well.

The price we pay for "free" services is to open the curtains that once hid our private sphere. If all the information I provide to various sites were kept isolated from other data, this would weaken my assumption that Big Data is a kind of assemblage that works as a functional entity. We know, however, that the different sites *do* sell the information we provide, and some major controversies have highlighted such problems, for example, the Cambridge Analytica scandal involving Facebook (Greenfield, 2018; Isaak & Hanna, 2018). Besides information that is provided voluntarily, information is also gathered from various other sources, such as GPS signals, satellite images, surveillance footage, information from mobile phone companies, smart electricity meters, etc. Some of this information requires consent, and some of it is gathered with government blessing. Zuboff (2019) and Véliz (2020) both provide comprehensive accounts of how data is being collected and used in modern society.

In sum, Big Data might turn into a form of Big Brother, in that it is *omniscient* and *ubiquitous*. While Big Data Brother *is* actually watching you, the political aspects of our societies and Orwell's Oceania are still quite different. Big Brother is the head of an imaginary totalitarian society, but this form of surveillance is not restricted to societies such as Orwell's Oceania. Big Data *may* take us there, though, and both Cohen (2012) and Berlin (2002) argue that the loss of privacy could lead to societies that are less liberal and less well-functioning.

Proposition 3.1[1]

Based on these considerations I have arrived at the first proposition:

> *Big Data in modern society involves gathering vast amounts of data, even the most private data, from individuals. Big Data thus threatens privacy and enables surveillance.*

Firstly, Big Data involves gathering large amounts of people's private data. It is possible to deny that this is happening or to state that Big Data is not *necessarily* connected to private data. I argue that it has been shown that it is in fact happening today, and it is this phenomenon I am discussing—not some ideal and limited alternative implementation of Big Data (Véliz, 2020; Zuboff, 2019).

Secondly, if massive amounts of private data are being gathered, privacy is, naturally, under threat.

Thirdly, the tools and routines used to gather the information described in this section *enable* surveillance. We may assert that the collection of data constitutes surveillance in and of itself, but we need not. It suffices to state that the data, when gathered, exist and *may* be used for surveillance purposes in some unknown future.

When Big Brother sees you

Privacy and private information

Privacy refers to the sphere in which one can expect to be completely unobserved, and it denotes both *information* and *acts* that are considered private. As such, privacy is a concept whose content can change depending on the user's views of what is properly private and public. It is a difficult term to define, and it has been used to express a confusing variety of concepts (Westin, 1967). Daniel Solove argues that privacy is a "concept in disarray" (Solove, 2005, p. 477; 2008, p. 1). It can refer to a wide array of slightly different concepts, including *the right to be let alone, limited access to the self, secrecy, control over personal information, personhood*, and *intimacy* (Solove, 2002). On a more positive note, Anita L. Allen (2011, p. 4) argues that privacy actually refers to a "predictable range of conditions and liberties," both for laypersons and experts. Rather than interpreting the wide array of possible meaning as proof of some sort of disarray, she sees this

[1]The propositions in the Chapters 3–5 are numbered X.Y, where X is the chapter number, and Y is the proposition number in the chapter.

diversity as a result of the fact that privacy is an "umbrella concept" (Allen, 2011, p. 60; Solove, 2005). Westin (1967, p. 7) proposed a definition that allows us to embark on the analysis:

> *Privacy is the claim of individuals, groups, or institutions to determine for themselves when, how, and to what extent information about them is communicated to others*

It is characterized by being *voluntary* and *temporary*, and it involves withdrawal from the public —either physically or psychologically (Westin, 1967). Privacy is thus a constant battle between my desire to withdraw and the curiosity of both the government and other people. The legal aspects of privacy are beyond the scope of this book, and I posit that people have a certain *right* to privacy (Warren & Brandeis, 1890).[2]

In order to provoke as few objections as possible, I will assume that this right protects only a minimal sphere that most people would agree should be private. This right entitles me to object when, for example, my neighbor wants to survey my house out of curiosity, or when the government has a desire to track my movements using a GPS tracker. There must be *some* sphere I can retreat into in order to meaningfully experience privacy—individuals must be able to deny the public access to certain parts of their lives.

Big Data threatens privacy in several ways. Firstly, if I live a standard life, using regular tools and services that are ubiquitous in our society, I will have to surrender private information. Secondly, even if I attempt to keep my strictly personal information to myself, many groups and organizations are now virtual, meaning that information about meetings, communication, etc., is gathered, and sometimes made public. This can also be construed as a violation of my privacy. Finally, some information is gathered against our wishes or without our knowledge, justified by concerns such as safety and security, anti-terror legislation, etc.

Big Data and superficial voluntariness

It may be objected that the first two threats are agreed to voluntarily, and that everyone has the option to refrain from such activities. One of the reasons why it is so hard to *not* feed into Big Data is that a lot of the services that collect *most* data are practically a necessity for people today. Social media sites such as Facebook are where friends and family communicate, colleagues

---

[2]I do not examine the notion of privacy as a *right* (or *normative privacy*) in depth in this chapter (Tavani & Moor, 2001). It is here granted tentative status as a right, and this will in turn be analyzed further in Chapters 6 and 7.

coordinate social events, and parents meet in various groups to coordinate their children's various activities. Not taking part is *possible* but comes at a high social price. Saying "no" to the user agreements provided by, for example, social media sites is of course *possible*, but may involve being "deprived of critical services" (Baruh & Popescu, 2017, p. 586).

For participation to be voluntary, alternatives must exist. When people are required to have accounts with, for example, Facebook, in order to get crucial information about their children's activities, studies, social events, etc., freedom of choice disappears. People are left with the choice between either grudgingly complying with the new normality of surveillance, or becoming like Luddites in a new social pariah caste. In addition, banks and government services are becoming increasingly digital, while traditional alternatives disappear.

As regards negative liberty, we do not have to agree with Thomas Hobbes that a person threatened with grave consequences is still at liberty to act freely. Hobbes relates the story of a man in trouble at sea who faces the dilemma of either (a) throwing his goods overboard to save himself or (b) sinking. According to Hobbes, the man is free to sink, so no liberty is lost (Hobbes, 1946, p. 137). In a similar vein, we are free to reject everything digital. *Voluntary* acts are *free* acts for Hobbes, but we rely on a broader set of theories of liberty. Within the framework of Berlin's negative liberty, for example, a robber with a gun obstructs and disrupts the liberty of the person he robs. His interference is clear, as his intention is to constrict our liberty and coerce us into acting in a certain way. The specifics of this debate are beyond the scope of this book, and I refer to Carter's (1999) discussion for more detail on how such situations do not necessary lead to a loss of *specific* freedom, while still leading to a loss of *overall* freedom.[3]

Foreshadowing one of the conclusions in this book, it seems likely that privacy is a good that is not best served by an individualized approach. I will argue that privacy is a public good, and that we must consider it a task for government to prevent situations where people are forced to abandon privacy in order to live what would be considered normal lives (Sætra, 2020a). I will also note that in the current setting, we do not need to say that privacy has intrinsic value. It may simply be valuable because it has *constitutive* value, in that it is a necessary condition for liberty (Carter, 1999).

---

[3]Lanier (2018) and Zuboff (2019) provide detailed accounts of the difficulties of escaping the gaze of Big Brother.

Proposition 3.2

Based on these considerations, I arrive at the second proposition:

> *A lack of alternatives to lifestyles that involve feeding into Big Data leads to something akin to forced participation in the surveillance of Big Brother.*

Freedom can be said to consist of having alternatives to choose from. In modern society, we are running out of alternatives that let us live *outside* the gaze of Big Brother. I fully recognize that theoretically we may avoid much of the surveillance. However, doing so involves incurring higher and higher costs because essential societal functions are organized through, for example, social media. Other parts of the surveillance are performed by government decree. Examples here would be the requirement that households install smart electricity meters in their homes or that all cars are required to have GPS transmitters.

Surveillance and freedom

Surveillance can be defined as "focused, systematic, and routine" *attention* aimed at obtaining some kind of information or attaining a purpose (Cohen, 2012; Lyon, 2007, p. 2007; Macnish, 2018; Wood, Ball, Lyon, Norris, & Raab, 2006). The *strategic* surveillance with a clear purpose is, however, not always the best fit for describing modern everyday surveillance with unclear intent and focus, and Macnish's (2018, p. 10) definition of surveillance as "sustained monitoring of a person or people" more easily allows us to see that the practices associated with Big Data are in fact surveillance. While the focus of such surveillance may be broad, there is still attention on each individual, and those behind the surveillance will be able to identify the actions of each observed person. Westin (1967, p. 57) distinguishes between *direct* and *indirect* surveillance. The first involves actively watching a particular person, whereas indirect surveillance involves gathering records and information that are subsequently used to determine the need for direct surveillance.

I distinguish between three kinds of surveillance. Firstly, it is important to note that surveillance does not necessarily lead to interference in people's lives. If so, *observation* is all this is, and the person is, technically, as free to *act* as they would have been without being observed. Someone might gather information about others without doing anything to steer, guide, or prevent the actions of the person observed. The electricity meters in our homes provide such information. Certain forms of video surveillance that are installed merely to observe and document behavior, etc., are of this kind. This I label *passive observation.*

Secondly, we can imagine that an act is committed under surveillance, and that the person observing uses the information as evidence in order to punish or reward the actor in retrospect. This form of surveillance could, of course, also be benign. Perhaps the information gathered is used to reward prosocial behavior, or in order to give good citizens credits in a government-run reward scheme. China is planning such a scheme whereby citizens will be given a "Citizen Score" that will determine their chances of obtaining loans, visas, etc. (Helbing et al., 2019). However, such schemes would strike most of us as totalitarian rather than benign, even if they were designed to provide rewards instead of punishment. While the first situation constituted passive observation, I call this *active observation.*

The third kind of surveillance is the one we most often think of, where those behind the surveillance use it actively to both punish and reward actions already performed *and* to prevent *planned actions.* If a government agency receives information that a group of individuals is planning a terrorist attack, it will intervene rather than use this information to punish the individuals afterward. This is *surveillance proper.*

The threat of passive surveillance

How do the three forms of surveillance affect liberty? To examine the situation that is *least* likely to be problematic, let us conduct a thought experiment. Assume that we have discovered a new form of being that we call the Observer. The Observer does not exist in physical space, has no memory, and has no possible means of communicating. It can, however, observe whatever it desires. Would we have cause to complain about being observed by this being that has no possible means of using, or abusing, the information it gathers and no possibility of influencing our lives in *any* other way than by the fact that it can observe us?

Imagine yourself at home, alone with your spouse. You have a desire for intimacy with your spouse, but suddenly you become aware that the Observer is watching you. Would this affect your behavior in any way? If we assume that its lack of memory and inability to store information is real, there is no reason to worry about the possible *abuse* of actions the Observer observes. I could, however, cite my *right* to privacy, and state that, for me, liberty partly consists in this right being protected. If the Observer sees me, I am not free, because my right to privacy is obstructed— purposefully—by another being. The *reason* I demand privacy is of little consequence, but I could cite reasons such as modesty and the desire to do things that would cause me to feel shameful if I knew that someone was observing me. Observation alone would hinder me. I could, for example, simply feel that the Observer would judge me. Being observed *changes* my behavior. Since I act differently when observed, I lack liberty if there is no space in which I can be unobserved.

The threat of active surveillance

When we can be punished after the fact based on evidence gathered, our actions are more likely to be affected by the observation. Some actions will be more costly due to the fact that I know that I will be punished. I do not take issue with liberty and the law, and take the view that liberty consists in living in accordance with a limited set of laws that are necessary to keep order. This is akin to Bastiat's (1998, p. 25) view that for an individual (proper) laws are not a violation of "his personality, his liberty, nor his property. They safeguard all of these."

Consider now my freedom to perform *immoral* acts—acts that people frown upon—without them being illegal. I would probably feel pressure to abstain from those as well. This is an issue Tocqueville (2004) notes in his examination of American democracy—the tyranny of the majority. The majority can create a set of opinions and actions that are accepted, while others are met with social sanctions without being illegal. In America, Tocqueville noted, this set was so limited, and the sanctions so harsh that people lacked both spiritual independence and *real* freedom of speech (Tocqueville, 2004). It is easy to envisage freedom of action suffering as a result of the same mechanisms as freedom of discussion.

The threat of surveillance proper

An interesting aspect of surveillance proper is that it leads to people being arrested for planning crimes not yet committed. Conspiring to do such things has been declared a crime, so a criminal act has been committed just by *planning* the actions, and acts of this kind are hard to uncover without surveillance of some sort. Thus, we have arrived at the most common justification for surveillance.

If we follow the argument of political philosophers like Thomas Hobbes, we accept that people will, by necessity, award *some* rights of surveillance to the sovereign power in order to preserve what is most important: safety, order, and the survival of the political community (Hobbes, 1946). While Hobbes states that the government should have no more such rights than *necessary*, he is not comfortable about erecting solid barriers the authorities may never pass. More liberal theorists, such as John Locke, do not have such qualms (Locke, 1969). Either way, we probably cannot both claim *absolute* rights against government surveillance *and* say that we want the government to ensure order. In order to use the concept of liberty in a meaningful sense, we must be able to consider ourselves free, even in society. This is a case where utility clearly trumps absolute liberty (Sætra, 2020b). But can we sacrifice liberty for other purposes than safety? This is the age-old question of political theory and the legitimacy of the state. According to Hobbes (1946), *security* and *survival* are the only things individuals can universally agree upon as goods. Thus, the state can legitimately limit our liberty through the use of surveillance aimed at promoting order, although this logic does *not* allow surveillance in order to promote other goods that might be seen as luxuries.

Danger of abuse

Finally, I briefly consider more traditional arguments against surveillance, such as (a) the possibility that the information will be used in other ways than we are led to believe, (b) the fact that information, being stored, may come into the wrong (or just *other*) hands later on, and (c) the possibility that the information will be used against us in new and novel ways in future that we cannot now foresee.

The first option involves either deception or simply a lack of understanding on the part of users. When people create their accounts on Facebook, they *assume* that the information they provide will not be used to target them politically, for example, in order to influence how they vote. One might argue that they *should* be aware of such risks. If sufficiently many do *not* understand the risks, government regulation might be necessary in order to prevent the exploitation of individuals' trust. This is partly because of a general desire to preserve privacy and liberty, but also partly because of the near mandatory nature of such services in modern society.

The second option involves factors such as the chance of human error; the sale and transfer of information between corporations, organizations, and government; and theft of data. The aspects of use, misuse, or abuse of data are given more attention in Chapter 6, as the *use* of information obtained is seen in conjunction with surveillance. While criminal exploitation of data, identity theft, hacking, etc., are certainly possible, a full examination of the degree of risk each of the above possibilities entails is beyond the scope of this book.

The third option is of great importance, since it suggests that we should adopt a precautionary stance toward privacy regulation. While providing data may not pose a risk *today*, we have no guarantee that it will not be used for malign (from our perspective) purposes *tomorrow*. What if new regulations allow such data to be shared and used for purposes such as pricing insurance, granting travel visas, gaining access to public services, etc.? And of particular relevance to the discussion in the next chapter: What if new technologies and insight allow us to use personal data to influence people more effectively in the future?

In addition to the potential threats of abuse, Big Data poses a threat in itself, so I might not even *need* to consider abuse in order to construct a liberal defense of privacy. When the foregoing points are combined with the possibility of abuse, however, we see the contours of an even greater threat to liberty.

Proposition 3.3

The above considerations lead to the third proposition:

> *Surveillance and the lack of privacy are a threat to freedom,*
> *because (a) the information gathered can be abused, (b) people*
> *have a right not to be observed (even if the surveillance is*
> *completely benign), and (c) being observed is an intervention that*
> *can affect the observed person.*

Not having privacy is a threat to freedom for three reasons. The first is based on a precautionary principle, which implies that surveillance and observation are wrong simply because they are *risky*—not because they are wrong in themselves. However, the second reason is that we can legitimately view it as wrong in principle to deprive people of a right to privacy. Thirdly. regardless of rights, surveillance is an intervention that changes my behavior and makes it costlier to perform actions that I would prefer to be unobserved. Some might say that someone who has nothing to hide should have no objection to surveillance, but this is a flawed argument. The costs created by surveillance apply to actions that are not illegal or constitute "something to hide" in the judicial sense.

Freedom under the gaze of Big Data

There are two main conclusions that follow from the premises presented in this chapter. They are (1) that Big Data poses a threat to both positive *and* negative liberty, but perhaps *especially* to the concept of liberty as nondomination and independence (List & Valentini, 2016; Pettit, 1997) and (2) individuals' liberty may be under threat even if the individuals themselves do not divulge information. Privacy may in fact be a public good (Sætra, 2020).

Big Data surveillance as a threat to liberty

The threat to positive liberty

The premises lead to the conclusion that positive liberty is under threat. As premise 3.1 is combined with 3.3, we see that a lack of privacy and being under surveillance threaten liberty both because of the risk of abuse of information and because the fact of being observed is a violation of a person's right to have a private sphere in which he will be neither observed nor disturbed.

While a negative conception of liberty may allow us to disregard many factors that *influence* a person because they are inimical to liberty, the positive conception does not. If I am to be my own master, I must have the privacy required to act as if unobstructed. Being observed is an obstruction insofar as it imposes various costs on actions that a person may have a desire to perform. If I am observed, and thus act differently than I would otherwise want to, my actions are not the actions of a positively free person.

It is possible to argue that, if I am truly my own *master*, I should be strong enough to withstand the pressure of expectation, and withstand the possibility of social sanctions. However, human traits such as modesty and shame must be accepted instead of wished away. This means that it is problematic to legislate on the assumption that people are, or *should be*, free from such influences. We should instead legislate to remove the possibility of making people less free by not taking things such as modesty into consideration. We might *wish* that our spouses, and others, had no problem with surveillance at home, so that the government could protect everyone even better, but since that is not the case, such a policy cannot be accepted.

Positive liberty requires that I be given space to both *become* and *act as* my own master, and surveillance can prevent both these things from happening. Berlin imagines an autonomous person "not acted upon by external nature or by other men as if I were a thing, or an animal, or a slave incapable or playing a human role, that is, of conceiving goals and policies of my own and realising them" (Berlin, 2002, p. 178). This, I argue, is not a person under surveillance.

The threat to negative liberty

For negative liberty, we can make a somewhat weaker—but still strong—argument for the case that Big Data is a threat. With the third premise, we may only need the first part, which concerns the risk of abuse, in order to argue that surveillance and lack of privacy are inimical to liberty. If so we would be constructing an argument that resembles Pettit's (1997) liberty as *nondomination* and List and Valentini's (2016) liberty as independence.

A practical doctrine aimed at protecting liberty in the negative sense requires safeguards against future interference, and I have shown that this will most likely involve restricting what sort of data can be gathered, or *at least* what information can be stored and transferred, and how it can be used—now *and* in the future. When nondomination is considered a prerequisite of liberty, I consider the mere fact that someone has power to coerce you and dominate you to be a violation of liberty, and not just active use of this force to control your actions. I propose an argument akin to Hobbes's view of war as the *possibility* of war, and not only active fighting, when I consider the possible threat to freedom posed by Big Data (Hobbes, 1946). The uncertainty that follows from the risk of data leaks, human error, hostile hacking attempts, the sale of data, change of ownership of companies, change of government, or simply a change in the intentions and plans of the actors we once trusted makes the mere existence of data a threat.

It can also be argued that being deprived of the right to privacy is in itself a violation of negative liberty, due to the combination of the first and second premises. Berlin (2002) discusses noninterference when discussing negative liberty. While being observed constitutes a violation of the space required to be positively free, I argue that it can also be interpreted as a form of interference.

Say that I, when at home alone, have a desire to watch movies that most people consider morally disgusting. If I can get a hold of such movies without anyone observing it, I would do so. However, if I know that others *are* registering and taking note of the movies I acquire, I might abstain. The introduction of observation is interference, and, if I cannot choose to not be observed, negative liberty is violated. I would argue that this was the case because (a) someone could use this information against me in the future, (b) I have a *right* to privacy with regard to my choice of movies, and (c) I dare not watch the movies I want to watch when I am being watched, so this observation interferes with my ability to perform lawful actions that hurt no-one. I could say that I am "coerced, or it may be, enslaved" because my inability to do what I desire is caused by the conscious and intentional actions of other people (Berlin, 2002, p. 169).

Conclusion

Big Data can be seen as constituting a form of surveillance. This surveillance is problematic, and freedom *is* threatened under the gaze of Big Brother. It is important to note that the technology of Big Data is not the key problem, and that it is the way we apply it that is threatening.

The three premises I have presented lead to the conclusion that all types of liberty are potentially threatened. There are therefore several reasons for adherents of liberalism to be wary of the effects of Big Data and how we currently regulate (or do *not* regulate) it, and in Chapter 6 I analyze the public goods aspects of privacy in more detail.

It could be argued that liberalism is concerned with freedom, and that this freedom means that businesses and government must be free to innovate and use new technology to gather data. As long as this is based on voluntary actions, many liberals would not object, and some would even claim that the *economic* benefits of how we use Big Data today more than outweigh the negative effects on privacy and other forms of liberty. In the areas of science and business, there are many examples of how the advent of Big Data has resulted in great benefits. I fully accept these benefits, but that does not alter the fact that it poses a threat to liberty. We must make a trade-off, and that means that we must consider how highly we value liberty.

I have begun to construct the argument for stating that liberty requires privacy, and that the individual's right to a sphere which no one can enter without express agreement is fundamental—more so than businesses' freedom to innovate and observe others. Big Brother's gaze must be averted, then, if people are to be free. Free to be their own masters, if you wish, or simply to ensure that people are not "coerced, and enslaved," to use the words of Isiah Berlin (2002, p. 169).

References

Allen, A. (2011). *Unpopular privacy: What must we hide?* Oxford: Oxford University Press.

Baruh, L., & Popescu, M. (2017). Big data analytics and the limits of privacy self-management. *New Media & Society, 19*(4), 579–596.

Bastiat, F. (1998). *The law.* Irvington-on-Hudson: Foundation for Economic Education.

Berlin, I. (2002). Two concepts of liberty. In H. Hardy (Ed.), *Liberty.* Oxford: Oxford University Press.

Böhme, R., & Köpsell, S. (2010). *Trained to accept? A field experiment on consent dialogs.* Paper presented at the proceedings of the SIGCHI conference on human factors in computing systems.

Carter, I. (1999). *A measure of freedom.* Oxford: Oxford University Press.

Cohen, J. E. (2012). What privacy is for. *Harvard Law Review, 126,* 1904.

Deleuze, G., & Guattari, F. (1988). *A thousand plateaus: Capitalism and schizophrenia.* Bloomsbury Publishing.

Greenfield, P. (March 26, 2018). *The Cambridge Analytica files: The story so far.* The Guardian. Retrieved from https://www.theguardian.com/news/2018/mar/26/the-cambridge-analytica-files-the-story-so-far.

Haggerty, K. D., & Ericson, R. V. (2000). The surveillant assemblage. *British Journal of Sociology, 51*(4), 605–622.

Helbing, D., Frey, B. S., Gigerenzer, G., Hafen, E., Hagner, M., Hofstetter, Y., … Zwitter, A. (2019). Will democracy survive big data and artificial intelligence?. In *Towards digital enlightenment* (pp. 73–98). Cham: Springer.

Hobbes, T. (1946). *Leviathan.* London: Basil Blackwell.

Isaak, J., & Hanna, M. J. (2018). User data privacy: Facebook, Cambridge Analytica, and privacy protection. *Computer, 51*(8), 56–59.

Lanier, J. (2018). *Ten arguments for deleting your social media accounts right now.* Random House.

List, C., & Valentini, L. (2016). Freedom as independence. *Ethics, 126*(4), 1043–1074. https://doi.org/10.1086/686006

Locke, J. (1969). *Two treatises of government.* New York: Hafner Publishing Company.

Lyon, D. (1994). *The electronic eye: The rise of surveillance society.* University of Minnesota Press.

Lyon, D. (2007). *Surveillance studies: An overview.* Polity.

Macnish, K. (2018). *The ethics of surveillance: An introduction.* Oxon: Routledge.

Marcuse, H. (2013). *One-dimensional man: Studies in the ideology of advanced industrial society.* Routledge.

Orwell, G. (1949). *1984.* New York: Harcourt.

Pettit, P. (1997). *Republicanism: A theory of freedom and government.* Clarendon Press.

Sætra, H. S. (2019). Freedom under the gaze of Big Brother: Preparing the grounds for a liberal defence of privacy in the era of Big Data. *Technology in Society, 58,* 101160.

Sætra, H. S. (2020a). Privacy as an aggregate public good. *Technology in Society,* 101422.

Sætra, H. S. (2020b). Toward a Hobbesian liberal democracy through a Maslowian hierarchy of needs. *The Humanistic Psychologist.*

Solove, D. J. (2002). Conceptualizing privacy. *California Law Review, 90,* 1087.

Solove, D. J. (2004). *The digital person: Technology and privacy in the information age* (Vol. 1). NYU Press.

Solove, D. J. (2005). A taxonomy of privacy. *University of Pennsylvania Law Review, 154,* 477.

Solove, D. J. (2008). *Understanding privacy*. Cambridge: Harvard University Press.

Tavani, H. T., & Moor, J. H. (2001). Privacy protection, control of information, and privacy-enhancing technologies. *ACM SIGCAS - Computers and Society, 31*(1), 6–11.

Tocqueville, A. D. (2004). *Democracy in America*. New York: The Library of America.

Véliz, C. (2020). *Privacy is power*. London: Bantam Press.

Warren, S. D., & Brandeis, L. D. (1890). The right to privacy. *Harvard Law Review*, 193–220.

Westin, A. F. (1967). *Privacy and freedom*. New York: IG Publishing.

Wood, D. M., Ball, K., Lyon, D., Norris, C., & Raab, C. (2006). *A report on the surveillance society*. UK: Surveillance Studies Network.

Zuboff, S. (2015). Big other: Surveillance capitalism and the prospects of an information civilization. *Journal of Information Technology, 30*(1), 75–89.

Zuboff, S. (2019). *The age of surveillance capitalism: The fight for a human future at the new frontier of power: Barack Obama's books of 2019*. New York: PublicAffairs.

4

Big Data nudging and liberty

Introduction

Imagine that I could make you do what I wanted you to do without you realizing that I was even involved. All I have to do is to rearrange the information around you in ways I know would lead you in the direction I desired. I could change the sequence of the choices you have to make, and use my knowledge of your susceptibilities and weaknesses to choose the appropriate time and method of delivering my *nudge*.

In this chapter, I examine how nudging powered by Big Data relates to liberty. I first examine nudging theory as expounded in Thaler and Sunstein (2008) classic book *Nudge*. The techniques of nudging are considered to be universal, as they can be used by both private and public actors, for any imaginable purpose (Sunstein, 2016a). I focus in particular on how liberty is affected by appeals to irrational mechanisms. I contrast nudging with *rational persuasion*, and, while some rhetoric and emotion is necessarily a part of rational persuasion, I consider it superior to methods of influence that appeal to our sub- or prerational faculties.

I then proceed to describe how nudging has *changed* with the advent of Big Data. Both the government and private companies gather data, and they use the data they have gathered to influence us in various ways. Nudging is based on knowledge of people and their susceptibilities, and I argue that nudging is becoming increasingly effective in three ways. Firstly, the fact that we now have more information about individuals means that we can nudge them more effectively. Secondly, given the amount of information we have about how human beings act, we now have increasingly sophisticated theories about how individuals function. They can be used to target human vulnerabilities in ever more effective ways. Thirdly, we now have the means to target people individually, through channels such as social media, online advertising, targeted location-based information on their phones, etc. Nudging is becoming more like precision bombing than the carpet-bombing of old, which makes it more effective.

Big Data's Threat to Liberty. https://doi.org/10.1016/B978-0-12-823806-6.00006-5
Copyright © 2021 Elsevier Inc. All rights reserved.

51

Schmidt and Engelen (2020) argue that nudging combined with Big Data (and AI and machine learning) is a key research area related to nudging, and this is precisely where I aim to contribute.

In sum, nudge comes to shove in the era of Big Data, and I argue that this development is inimical to liberty. I will develop a set of propositions that will subsequently be used as premises for the argument that nudging may constitute a form of secret coercion that is deeply troubling, and that we should not use the difficulty of regulating it as an excuse for not trying to do so. Nudging has given rise to principled objections since its inception and, as nudge comes to shove, it becomes coercive and a threat to liberty.

Nudging

"Nudging" is a term used to describe an approach to behavioral modification, without the use of force, by both private and public actors.[1] The term was coined by Cass R. Sunstein and Richard H. Thaler, who argue that it is not only possible, but also legitimate, to "actively influence" behavior. Furthermore, they argue that this can be done while respecting people's freedom (Thaler & Sunstein, 2003, p. 1). The art of influencing behavior is not new, but the new combination of insight from behavioral economics, cognitive psychology, and social psychology made Sunstein and Thaler's theory an important addition to existing knowledge (Goodwin, 2012; Thaler & Sunstein, 2003).

They call their approach *libertarian paternalism*, because the *goal* we nudge people toward should be people's own *welfare* (Thaler & Sunstein, 2003). The issue of who is the best judge of what *my* welfare is will have to rest until later. For now, I will simply point out that idealism is all well and good, but the tools of nudging could just as easily be used for nonidealistic as for good purposes (Mills, 2020a). When we take what some would call a realistic—not idealistic—stance on motivation, the theory becomes less paternalistic and, quite simply, manipulative (Sattarov, 2019). People's actions are steered toward *whatever* goals those who nudge us have—also toward what Sunstein labels *illicit goals* (Sunstein, 2016a).

Nudging arose from an understanding of *irrationality*—the fact that people often make "bad" choices. Too bad for rational

---

[1]Various authors have developed concepts related to nudge, such as *sludge* and *budge* (Mills, 2020a; Oliver, 2013, 2015; Thaler, 2018). I will not distinguish between these forms of behavioral modification in this book, and treat them all as forms of nudges.

choice theorists, who rely on the "false assumption" that people usually make choices that are in their best interest (Thaler & Sunstein, 2003, pp. 4, 9). *Real* people have cognitive weaknesses. Thaler and Sunstein (2003, p. 9) refer to such things as us failing to live up to Bayes' rule, that we use rules of thumb "that lead us to make systematic blunders," that we sometimes "prefer A to B *and* B to A," that we lack self-control, and that we are influenced by the framing of the issues we face.

The idea is that people often lack both self-control and a proper long-term perspective. This leads them toward choices that, among other things, make them obese and poor. This is because saving and healthy eating are examples of behaviors that require both a long-term perspective and some self-control (Thaler & Sunstein, 2003). We make better choices in situations in which we have "experience and good information," which means that we are better at choosing the "correct" ice cream than we are at investing our money wisely or choosing the most beneficial medical treatment (Thaler & Sunstein, 2003, p. 5). The latter actions are "complex, uncertain, or otherwise challenging," and in such situations, many do what is cognitively easier than reaching what some categorize as the *best* decisions (Dotson, 2012, p. 328).

To be slightly technical, there are several reasons for our bad choices, such as the *stickiness of the default option, anchors,* and *framing effects* (Thaler & Sunstein, 2003). People tend to choose the *default option* when they have no clear basis on choosing between alternatives, and *anchors* similarly focus people's attention on some alternative that is presented at the beginning of a choice process (Thaler & Sunstein, 2003). How something is portrayed affects how we interpret different alternatives, and this is what we refer to as *framing effects* (Thaler & Sunstein, 2003). We also often lack clearly formed preferences, and they even change depending on how we choose—our preferences are "ill-formed and murky" (Thaler & Sunstein, 2003, pp. 6, 23). Another factor is *suggestion*: when uncertain, we rely on (a) what most people do or (b) what experts do. Yet another explanation is *inertia*, which implies that moving away from the status quo is costly in some way, and is therefore often avoided. Lastly, we have the *endowment effect*, which points to the fact that people "value goods more highly if those goods have been initially allocated to them" (Thaler & Sunstein, 2003, pp. 22–23).

Proposition 4.1

These considerations lead to the first proposition:

Nudging theory is based on the disciplines of psychology and behavioral economics, which deal with what affects people's choices, irrational choices included. It has proven to be effective, and the insights from the theory can be used by anyone with access to knowledge about people and how they act, for whatever purposes.

Firstly, nudging theory is described as a theory about *influencing* people's choices, and it is based on disciplines that deal with understanding how human beings think, act, and decide (Thaler & Sunstein, 2003).

Secondly, many experiments, both in laboratories and real-life implementations, show that the techniques of nudging have real effects on people's behavior (Sunstein, 2016a). This is not to claim that it is *always* effective or that it is *fully* effective when employed. Some nudges fail, as Sunstein (2016b) highlights in *Nudges That fail*, and, even when they work, they do not determine every individual's actions. They merely change the *proportion* of people who make the choices they are nudged toward.

Finally, the techniques of nudging are available to all actors, regardless of their intentions and goals. This is merely to state that while some claim that nudging should only be used for good purposes, it can just as easily be used by business owners to maximize profits, etc.

When the nudge is powered by Big Data

The rise of Big Data

As I explained in Chapter 2, this is not a detailed technical examination of Big Data. For my purposes it suffices to state that, with Big Data and the associated means of tailoring information to individuals, (a) we get more information about individuals, (b) we get more information about how people function in general, and (c) we have the means to deliver nudges to specific individuals based on our existing knowledge of these people's preferences and inclinations.

Furthermore, Big Data is not just something private companies are involved in. Government (a) gathers data and (b) "is an important secondary beneficiary" of the data gathered by others (Cohen, 2012, p. 1916). As we saw in the previous chapter, the combined effect of private and public efforts to gather data can be referred to as the "surveillant assemblage" (Haggerty & Ericson, 2000). Although private and public gathering of data for different purposes *can* be considered as isolated phenomena, so much data flows in so many directions that I argue that a holistic view of the phenomenon is justified.

Richards and King (2013) point to certain paradoxes of Big Data, two of which are of particular interest in this context: those of *transparency* and *identity*. The first relates to the fact that the collection of information is often hidden, and that "its tools and techniques are opaque" (Richards & King, 2013, p. 42). The second concerns identity formation and, more specifically, how Big Data enables actors to "use information to nudge, to persuade, to influence, and even to restrict our identities" (Richards & King, 2013, p. 44).

When Big Data powers the nudge and makes it a shove

What is new about Big Data is (a) the combination of insight into the mechanisms behind decision-making and (b) highly detailed knowledge of personality profiles. Advertisers and a range of other actors have attempted to manipulate our decisions way before behavioral modification was branded as nudging (Lepore, 2020; Packard, 1957), but the tools now available to them mean that people can be nudged more effectively than before.

While writing this, I noticed an advertorial from an online casino, which serves as a good example of this phenomenon. It boasts about its innovative nature, and how Big Data and the individual tailoring of content enable companies to attract *more* clients who play *more* and stick around *longer*. With the data now available, the casinos know more about what games each player likes, the kinds of odds and bets they like to play, and what makes them play, stop playing, and change games in the first place. Even the *clients* are said to appreciate this, as they get exactly what they prefer and believe they desire (Kunstig intelligens, 2020). They might be happier for a while, but the ones who lose all their money will probably be left wondering what on earth happened, as they are thrown off this dazzling carousel ride that felt like just what they wanted.

Knowledge of your propensities lets actors lead you. You might have certain inclinations that you are not even consciously aware of, but that the data about you reveal (Baruh & Popescu, 2017). Being led in this manner is not necessarily experienced as oppressive. Skilful marketing based on these principles is rarely even noticed by the consumer, and people might feel freer than ever. While we previously had "predetermined and inevitably artificial categories" as targets of marketing and policy, we now get information and predictions "finely tailored to particular situations"—and individuals (Calo, 2013; Cohen, 2012, p. 1921; Mills, 2020b).

The concept of Big Data-powered nudging has yet to receive systematic academic attention, but Yeung's (2017) article "'Hypernudge': Big Data as a Mode of Regulation by Design" discusses some of the topics covered in this chapter. Yeung argues that "Big Data nudges are extremely powerful and potent due to their networked, continuously updated, dynamic and pervasive nature (hence 'hypernudge'),'" and she claims to be providing a "liberal, rights-based critique" of this phenomenon (Yeung, 2017, p. 118).

Yeung deals with the "liberal manipulation" critique of nudging, consisting of (a) the questioning of illegitimate motives and (b) the proposition that nudging can be construed as deception (Yeung, 2017, pp. 123–124). I argue that her critique of the deceptive qualities of nudging is not particularly effective, since it seems to rely on a person's "right not to be deceived, rooted in a moral agent's basic right to be treated with dignity and respect" (Yeung, 2017, p. 127). While this may be a fine goal, a liberal need not go this far and can instead argue, as I do, that "hypernudging" may be illegitimate simply because it deprives people of their *liberty*.

Helbing et al. (2019) speak of the "big nudge" when Big Data is combined with nudging. They point to its effectiveness, but also to various problems, such as its potential for abuse. The abusers could be criminals, or perhaps foreign powers nudging people politically, for example, in order to interfere with elections. It could also be hidden government activity—activity that is not desired or approved of by the people. This is one of the main problems for proponents of nudging: the problem of determining what is considered *proper* use of such techniques.

Outside of academia, the combination of Big Data and nudging techniques has been seen as both dangerous and potentially enormously beneficial. What the articles referred to in the next paragraph show is an appreciation of how Big Data combined with nudging creates a phenomenon of greater power, and thus greater potential for both good and bad, than nudging as we used to know it.

Guszcza (2015) discusses how Big Data and nudging can solve the "'last mile-problem'"—when you know where you want (others) to go, but can't get there. Eggers, Guszcza, and Greene (2017) talk of a "supercharged" nudge, and how "big data and the Internet of Things" offer opportunities for improving the effectiveness of nudging, and "improving government." Coughlin

(2017) warns us that we can "take nudge theory too far" with the help of Big Data and the Internet of Things. The shortcomings of traditional nudging are overcome by using Big Data to "better understand the real-time mood of the person, how much digital noise they are willing to tolerate or how much nudge noise they are being subjected to at that very moment" (Coughlin, 2017). Kittur (2020) writes of how *Big Data "nudges" lead to better merchandise decisions*, and refers to Amazon's Jeff Bezos, who states that their "Selling Coach" program generates over 70 million automated machine-learned nudges a week that "translate to billions in increased sales to sellers."

In politics, too, nudging is being taken "further than ever before," with governments creating "behavioural units," like Britain's "Behavioural Insights Team," which is working to create the best nudges with the help of Big Data (Giorgione, 2017; Sunstein, 2016a). Hugill (2020) also describes the Behavioral Insights Team, and how it has added data science "which aims to use the latest methods from data science, machine learning and predictive analytics to make smarter policy implementations." Such developments should leave no doubt about the importance of discussing the legitimacy and implications of nudging.

To sum up: Nudging is "rooted in an understanding of how people actually think" and this understanding is becoming much more solid in the age of Big Data (Thaler & Sunstein, 2003, pp. 23–24). Furthermore, "[p]resentation makes a great deal of difference" (Thaler & Sunstein, 2003, p. 24). With Big Data and social media, presentation can be tailored to fit individuals, as can the *delivery* of the nudges (Mills, 2020b).

The data may, for example, show that I am most susceptible to nudges toward impulsive purchasing decisions in the morning, whereas the sweet spot for targeting you is in the evening. In addition, the data may show that I am most susceptible to subtle nudges that streamline my choices and appeal to my emotions, while you are more easily swayed by anchors and framing effects properly deployed. Today, the nudgers can target both of us in the optimal way, whereas they previously had to either choose *one* of the approaches or some sort of compromise.

Proposition 4.2

Based on the considerations above, I arrive at the second proposition regarding the combination of nudging theory and Big Data:

> *With Big Data, nudging can become so effective that it is hard to withstand it, making the nudge more of a shove.*

As described in this section, Big Data provides so much detailed information about both (a) individuals and (b) the way people act *in general* that nudging becomes increasingly effective. In addition, (c) both the content and delivery can be tailored to individuals. While a nudge can be considered a welcome and slight push in a certain direction, a nudge fueled by Big Data can easily become more of a solid push, or shove, that is hard to withstand.

Nudging and liberty

The distinction between negative and positive liberty is at times used in the debate about how nudging affects liberty (Goodwin, 2012; Hausman & Welch, 2010; Mills, 2013). I argue that nudging is in conflict with *both* conceptions of liberty and the requirements of liberty as nondomination, independence, and autonomy (List & Valentini, 2016; Pettit, 1997; Raz, 1986). In the following, I mainly discuss the liberty reducing nature of nudging by discussing the aforementioned previous contributions to the debate about nudging. While they are aimed at nudging in general, they are equally relevant for Big Data-based nudging, which I argue to be even more problematic than the traditional type.

Is nudging liberty reducing?

The debate about nudging in politics is now over 10 years old, and I will briefly discuss some of the main points in the debate that followed the popular book by Thaler and Sunstein (2008). I do not limit my usage of the term to government nudging, and argue, along with Sunstein (2016a), that nudging can be employed by anyone, for whatever purpose. For example, advertisers nudge, and they have been doing so for a long time (Goodwin, 2012; Hausman & Welch, 2010; Packard, 1957).

Nudge marketing is a concept that builds on the same principles as government nudging (apart from the paternalistic demands imposed at times by Sunstein and Thaler) (Dholakia, 2016; Entis, 2014). I treat private and public nudging as equal with regard to their impact on liberty.

The difficulty of regulating private nudging should not preclude a debate about its implications for liberty. Nor should it imply that we had better resign and allow it, or even promote *government* nudges, simply because private actors nudge. Some, like Hausman and Welch (2010), seem to suggest that private nudges are acceptable due to their ubiquitous nature and resultant lack of effectiveness (Schmidt & Engelen, 2020; Sunstein, 2015). I argue that the difficulty of regulating nudging is not an excuse for allowing it, and I also argue that nudges are becoming more and more effective. In effect, these arguments in favor of nudging are considered both weak and unsatisfying.

Nudging as I define it is an attempt to influence an individual's decision-making by appealing to subconscious mechanisms or known irrational proclivities. I contrast this with *rational persuasion*, which consists of attempts to influence behavior through open and transparent appeals to reason. I regard nudging as troubling, and I argue that it is more so today than at any time before, because of new technology and the information we now have about both human actions in general and individuals. If traditional nudging was perceived as "deeply troubling," nudging powered by Big Data is arguably terrifying (Goodwin, 2012, p. 86).

Negative liberty

Goodwin (2012) states that the concept of nudging is based on the idea of negative liberty, and that this leads to problems when *positive* liberty is considered as well. For him, nudging is compatible with negative liberty, since it preserves *freedom of choice*. Hausman and Welch (2010, p. 124) state that Sunstein and Thaler understand liberty as the "absence of obstacles that close off possible choices or make them more costly in time, inconvenience, unpleasantness, and so forth." In their article, they employ the definition of liberty that Sunstein and Thaler subscribe to, which is quite close to the conception of negative liberty we have already established (Hausman & Welch, 2010). However, I argue that Goodwin is not correct in either of the assumptions above, as (a) negative liberty is more than he makes it out to be, and (b) negative liberty can be negatively affected by nudging. Mills (2013) also criticizes Goodwin for having an overly simplistic view of some key concepts, such as paternalism and autonomy, and I argue that this also applies to his idea of negative liberty.

What he presents as negative liberty is a rather Hobbesian concept, where even a man who is robbed at gunpoint is free, because he has the freedom to choose to resist the robber. For Goodwin, "negative freedom is merely concerned with whether you are, strictly speaking, able to make a choice" (Goodwin, 2012, p. 88). Furthermore, it requires not being *physically* obstructed by external obstacles (Goodwin, 2012, p. 88). This exact example is often used to describe the Hobbesian notion of liberty, and not Berlin's (Watkins, 1967). This is akin to the Hobbesian sailor who is free to sink if he dislikes the only other option, which is to abandon his goods in order to save himself and his ship (Hobbes, 1946). When I speak of negative liberty, I will not just consider *physical* obstacles, and I do not consider *theoretical*, but very *costly* alternatives sufficient to say that freedom is preserved. If another person, by conscious actions, makes it exceedingly costly for me to do what I might otherwise do without hurting anyone, or without them having other reasons that makes their actions reasonable, this person restricts my liberty in a negative sense.[2] So, while nudging could be unproblematic based on Goodwin's (2012) understanding of liberty, I object to this understanding, and will not employ it in the following.

Hausman and Welch (2010) argue that, while they do not think nudging is *libertarian*, they do agree with Sunstein and Thaler that it would (a) not be coercive, and (b) "not significantly" limit freedom of choice (Hausman & Welch, 2010, p. 124). I disagree with them on the first point, and, as to the second, it all depends on what one considers a *significant* limitation of freedom of choice.

However, nudging surely makes certain actions more *costly*, *inconvenient*, and *unpleasant*, and could thus constitute an obstacle to liberty. It is what Mills (2020a) labels *hedonic* costs, which is one of several noneconomic costs that are permissible under nudge theory. This, I suppose, is where Hausman and Welch (2010) would say that the unpleasantness caused by requiring producers to print horrible photos meant to discourage smokers from smoking is not *significantly* limiting their freedom, even though it will surely cause displeasure and inconvenience.

Thaler and Sunstein propose that their libertarian paternalism of nudges does not constitute *mandates*, and that people will still be free to smoke, eat badly, and invest poorly (Goodwin, 2012, p. 88). It is also important to note that Sunstein (2016a) clearly argues from a utilitarian stance: nudges may be theoretically problematic, but still worth it. This also leads him to warn repeatedly of the "trap of abstraction, which can create serious confusion" (Sunstein, 2016a, p. 26).

I will pursue the implications for the kind of liberty I have here described when discussing the coercive nature of nudging. In contrast to Goodwin (2012, p. 88), who believes that champions of negative liberty consider it an "abuse of words" to see false consciousness, psychological pressure, lack of awareness, etc., as obstacles to liberty, I keep this door open. This relates to the efforts in social theory to understand the various forms of *power* (Sattarov, 2019)—efforts that nudge theorists would do well to examine in more depth. Not only *physical* obstacles to action are considered threats to negative liberty, and a hypothetical

---

[2]I will not consider the question of how the law relates to liberty, and arguments such as Bastiat's proposition that the prohibition of unlawful acts is in accordance with liberty, since such prohibitions, through law, are preservers of, rather than threats to, liberty (Bastiat, 1998, p. 25).

freedom to choose will not necessarily be considered sufficient to render a person free if the set of choices is limited, manipulated, and associated with artificially high costs of various sorts.

Positive liberty

One of the points commonly discussed in relation to nudging is *paternalism*. Nudging is often condoned because it is done in the interest of the person being nudged. I will not focus on this aspect of Sunstein and Thaler's theory for two reasons. Firstly, I consider nudging to be a universal phenomenon that is not only associated with the government and good intentions. Secondly, I believe that it is excruciatingly difficult to arrive at a theory that leads to acceptance of the idea that one person can force, or mislead, another person to act in ways he does not wish to act, or at least does not know that he wishes to act, without his approval. That government can ban certain actions, and make undesirable actions costlier, is all well and good, as long as this is fully transparent and a matter of politics and social choice. While Sunstein and Thaler argue in favor of a similar idea of transparency, it seems obvious that affecting choice by appealing to subconscious and irrational mechanisms is *not* the same as overt governmental paternalism.

Let us assume that we have different ideas about what would be the best course of action in some imagined setting. All my attempts at rational persuasion fail, but I am *still* convinced that I know better than you do, and I want to enforce my will. I only have three options, which are (a) to overtly coerce you, (b) to entice you by external rewards, or (c) I could exploit my knowledge of the imperfections of your decision-making processes to make you act in the way I want you to (Hausman & Welch, 2010, p. 126). How is the last option more acceptable than the two that are based on open and transparent incentives? If we somehow argue that it is acceptable because you do not *know* that I have achieved my goal, a whole array of mechanisms of manipulation, covert coercion, and subterfuge will be available to influencers of various sorts, and I find it hard to believe that supporters of nudging really want to lead us to that conclusion. Furthermore, the popular support for nudges emphasized by Sunstein (2016a) does not affect the issues I raise with regard to liberty.

I follow Hausman and Welch (2010, p. 128) when they argue that a person's autonomy is diminished when "pushing" becomes more than rational persuasion. When a person chooses what they do because someone has actively interfered with the available choices and the presentation of alternatives, this must surely be considered a violation of a person's right to be their own master in a broad sense. It would violate the demands of positive liberty, nondomination, and varieties such as Raz's autonomy-based liberty with clear requirements of independence (Berlin, 2002; Pettit, 1997; Raz, 1986).

Sunstein discusses the relationship between autonomy and nudging, and seems to argue that people delegate certain decisions to the government, and that this does not infringe on autonomy (Sunstein, 2016a). However, the argument that *some* arrangement of choices would have influenced him either way is not a weighty argument in favor of allowing active manipulation in order to achieve someone *else's* goals through subterfuge.

Let us assume that you are blind, and that I somehow got into your apartment before you got up and got ready for work one morning. I saw that you had laid out your outfit for the day, and I found the color palette you had chosen utterly distasteful. I then quickly went out and bought clothes that were much more stylish, and put them in place of what you had laid out. I put the original outfit back in your closet, so that it was still available. I then left. You got up, put on the clothes I had put there, and went to work. I got you to wear what I wanted without you even knowing that I had done so, and I did it all for your own good, as I consider my sense of taste to be better than yours. Particularly since you are blind, I might feebly argue, if you were to challenge me on this, when your coworkers ask you why you are suddenly all colourful and trendy. Yes, you would have worn clothes anyway, but that is in no way an argument for me to make you wear something *else*. No matter how distasteful, random, or unfortunate your choice would otherwise have been. I could of course have woken you up and asked you if you wanted some help, but I could not nudge you toward trendiness in the way described here. That would be akin to changing the default option, which is one of the most common tools of nudging.

In this sense, the ideals of nudging, which supposedly only involve marginal interference with liberty through the arrangement of choices, etc., mean that a person's power to choose, and thus his autonomy, is interfered with (Hausman & Welch, 2010). Hausman and Welch (2010, p. 135) state the obvious when expressing their concern that "exploiting decision-making foibles will ultimately diminish people's autonomous decisions-making capabilities."

But can a nudge not *improve* people's positive liberty, and empower them? If irrationality makes me act in ways I would not approve of if I understood them, then perhaps I am *freer* in a world where my choices are laid out in ways that minimize the unfortunate effects of my cognitive weaknesses and lack of will power. For example, apps and various services might nudge people to decrease time spent on screens and various social networks. Mandatory cooling-off periods before making choices that we know people often regret, for example, could be seen as empowering in this way (Hausman & Welch, 2010, p. 126). It is equivalent to a policy of forcefully tying people to the mast in order to resist the siren song.

However, nudges are not always mandatory and transparent. When they *are*, I argue that they are more akin to regular government regulation based on traditional paternalistic principles, and should be viewed as such. According to C. Mills (2013, p. 29), nudges are better than traditional regulation because they preserve choice, but his evaluation is dependent on his qualification that nudges lead us to actions that we "would not disagree with." If I am not given a choice, or am not informed of what is being done, I find it hard to see this form of guidance as more liberty friendly than regular overt regulation.

Nudges, C. Mills (2013, p. 29) states, may lead us toward an "authentic life," by letting us overcome our irrational impulses. He argues that positive liberty should be concerned with "authentic decision-making," whereas "biases, blunders and temptations" are obstacles to this (Mills, 2013, p. 30). If so, helping us overcome them is clearly empowering, as it lets us achieve our "authentic" goals (Mills, 2013, p. 30). This line of reasoning is analogous to the debate on paternalism, where one opens the door for someone to label a person's goals and actions as *inauthentic* and in need of correction.

Goodwin (2012) refers to Taylor and the idea that positive liberty is about individual independence and self-fulfillment. He then seems to argue that nudging probably *cannot* be empowering, because nudging is purportedly based on the Hobbesian liberty he called negative liberty (Goodwin, 2012). This is naturally a fallacy, as a theory may perfectly well be compatible with concepts it does not explicitly discuss. A theory *may* be empowering, even if it is founded on the idea of negative liberty.

In addition to empowering through freedom from irrationality, nudges can empower by providing freedom *not* to choose, freedom from clutter, etc. (Mills, 2013). Sunstein and Thaler rely on a specific form of rationalism when they argue that what they label *irrational* interferences in our behavior should be corrected, but I cannot deal fully with the concept of rationality here (Hausman & Welch, 2010, p. 126).

Sunstein and Thaler prefer nudges that are transparent and subject to monitoring and acquiescence (Mills, 2013). However, they are also clearly aware that nudges are most needed, and most effective, when people face complex issues and have a hard time understanding what they really want, and what actions lead to what outcomes (Mills, 2013). Transparency and voluntariness are hard to achieve when the issues in question are in fact too complex to grasp. Otherwise, rational persuasion would most likely suffice, and I find it hard to believe that anyone would *prefer* nudging to rational persuasion if they were equally effective and available. We will also see that nudges are most effective when they are *not* transparent. If, as C. Mills (2013, pp. 30–31) states, personal autonomy is based on individual self-control, "reflective authenticity," and "independence from coercive and manipulative influences" it is difficult to accept the view that nudging is conducive to this form of liberty, or any other.

The manipulative and coercive nature of nudging

Nudging and coercion

One of the defining characteristics of liberalism is an aversion to coercion. Particularly the proponents of negative liberty cannot see coercion, over and above what is necessary for society and order to exist, as compatible with their idea of liberty. But what about manipulation, or even secret coercion? Raz (1986, pp. 377–378) distinguishes between *manipulation* and *coercion* by stating that the first does not alter a person's options, but instead "perverts the way that person reaches decisions, forms preferences or adopts goals." This involves the use of *episodic power*, in which actor A has a context-dependent power to, for example, influence or manipulate actor B (Sattarov, 2019). I argue that by, for example, manipulating choice architecture and framing, nudging might also involve interference with options, which thus makes it coercive, according to Raz (1986).

Let us assume that there is a mountain between where I *am* and where I *want to be*. I cannot complain that I lack the *liberty* to get there quickly because of this obstacle, unless I want freedom from *necessity* and *nature*, which is not a kind of liberty I am interested in here. While Sunstein (2016a) claims that *nature* nudges, I argue that it does not, and that it is the morally attributable actions of others that are of interest. But what if *you* stood between me and where I wanted to be, and physically pushed me back and stopped me each time I attempted to pass you? You would be exerting force on me, and my liberty would suffer as a result. Thus far most would likely agree on what constitutes liberty reducing force or not.

However, what if, instead of physically stopping me, you let me know that you would shoot me dead if I attempted to pass. This is where Goodwin (2012) mistakenly believes that the champions of negative liberty would feel that liberty was preserved. Just because you do not use *physical* force to change my actions, you are stopping me just the same if your threat of force is credible. Carter (1999) calls these coercive threats, but he also demands that the threats are actually enacted for them to be coercive. However, people act on the *belief* that the threat will be enacted, and I thus use a less strict definition of such threats. You have a desire to prevent me from doing something, and you impose a very high cost on the action you wish to prevent. While I still have the *choice* of being shot dead, I cannot be said to be free when you have intentionally changed my incentives in such a manner. If you inform me of your intention, my actions will be changed, but what if you do not let me know that you intend to shoot me? *Then* I would be as free to act as ever, and I would attempt to go there. You would probably shoot me, and then I would be a victim of murder, but not really *unfree*.

Social sanctions might also be relevant in this context. Say that you really wanted me to stay away from where I was going, and had a desire to persuade me not to go. If you provided good reasons that made me change my mind, I argue that my freedom would not suffer. This position necessitates a distinction between certain conceptualizations of power and their effects on liberty. While the power to influence is indeed a form of episodic power (Sattarov, 2019), it does not necessarily follow that it is inimical to liberty.

But what if you provided various incentives to affect my actions? If you offered to pay me 100 dollars to stay away, and I valued those 100 dollars enough to do as you wish, I would be perfectly free, as I simply used my freedom to make a transaction and changed my intentions. But what if the incentive was *negative?* If you broke the law and promised the penalty of death for my action, you would be using illegitimate force to compel me to stay away, and my freedom would be limited as a result.

But what if you simply said that you would hate me forever and never talk to me again? While I might value your friendship and decide *not* to go because of this, I would of course not be deprived of my liberty in this process. You are free to exercise your right to like or dislike my actions, and I have no cause to complain about loss of liberty.

However, what if you built a wall between me and my goal? You did this before I arrived, and you left before I got there, so I merely saw the wall when I arrived—much like how we face personalized web pages on sites such as Amazon, without really seeing anyone rearranging the pages in order to influence us. Imagine that you even made the wall look exactly like a mountain, seemingly the same obstacle that we considered earlier. Would this be an obstacle that deprived me of liberty? I argue that this is akin to the nudge, and I consider it to be a violation of liberty. The reason for this is that it has been constructed with the intention of changing my actions, and it conflicts with the negative requirement not to be interfered with by other people, the positive requirement to be autonomous and master of one's own actions, and the demands of nondomination and independence.

Rational persuasion

Rational persuasion is used as a contrast to nudging, and I consider rational persuasion to be fully legitimate. Such attempts at persuasion protect individual liberty and let agents retain control of their own goals and actions (Hausman & Welch, 2010; Mills, 2013). This is in strong contrast to manipulative efforts to change an individual's actions by exploiting their cognitive weaknesses and "circumventing the individual's will" (Hausman & Welch, 2010, p. 129). The difference between attempting to persuade someone and taking advantage of irrational mechanisms is important (Hausman & Welch, 2010). It is also interesting to note that Sunstein (2016a) labels efforts to *inform* as nudges, while I see this as rational persuasion.

Furthermore, the *goal* of the deception is of little interest. The consequentialist ethic, as exemplified by Sunstein (2016a), might justify nudging because the good effects it produces may be legitimate, but it would still be the case that liberty would suffer from it. We can, of course, argue that we value good effects more than liberty, and, if that is what the nudgers argue, all is good, and we can choose the politicians who propound the view we are most content with.

Manipulation and secret coercion

My contention is that any liberal who is opposed to *overt* coercion should not be indifferent to *secret* coercion. While the overt kind gets most attention, nudging combined with Big Data makes it possible to guide people's decisions in a way that is no less problematic than other forms of subtle guidance. Nudging is, in principle, inimical to liberty, but I claim that it becomes more problematic the more effective it is, and particularly the more *covert* it is. Goodwin (2012, p. 91) calls for more theorizing about the manipulative nature of the nudge, especially since the "nudge's libertarian credentials are undermined by the fact that it targets individuals in their pre-rational state." It is this aspect of nudging I take issue with, and I have clearly distinguished it from *rational persuasion* and traditional overt regulation using incentives.

While Hausman and Welch (2010, p. 124) argue that nudging is not coercive, they recognize that it could be a problem that a person's actions "reflect the tactics of the choice architect rather than exclusively their own evaluation of alternatives." They take a rather lenient stance on nudging, given that they clearly see that, when nudges aim to undermine an "individual's control over her own deliberation, as well as her ability to assess for herself her alternatives, they are prima facie as threatening to liberty, broadly understood, as overt coercion" (Hausman & Welch, 2010, p. 130). Goodwin (2012, p. 89) also sees the same attempt to undermine control of deliberation as a cause for concern. For C. Mills (2013, p. 32), what makes nudges troubling is that they "override and circumvent the autonomous agent's rational decision-making capacities," which thereby diminishes the agent's autonomy. He then calls this a form of manipulation of moral concern (Mills, 2013).

I argue that secret, or covert, coercion, is as much of a threat to liberty as the more obvious kind. I would argue that it is even more of a threat (a) as it is difficult to perceive and (b) since we do not have the same instinctual reaction to it as when someone *physically* coerces us.

Take the example of subliminal messages (Hausman & Welch, 2010). If they were effective, and if they could be used for beneficial purposes, would people still be free when influenced by them? In our previous example, let us say that you gained control over my television set and, for the past few weeks, you have been flashing rapid alternate images of death and devastation mixed in with images of the location I wanted to go to. All of a sudden I no longer want to go there, but I have no idea why. I just do not feel like it anymore. The mere prospect of going there makes me slightly queasy. No need to shoot me or block me any longer, as you have achieved your goal by using *psychological* force.

Am I still free, even though you have achieved your goal and made me to change my actions—or indeed my very goals—by your actions? Or is it really the case that overt coercion, for example, a requirement to use safety belts in cars on threat of punishment, is less of a threat to liberty than the covert kind? Could it be that an unhappily coerced person is in fact freer than a happy, but unknowingly manipulated, person is (Hausman & Welch, 2010)? I say yes—while ignorance can surely be bliss, it does not promote freedom. If it does, we'd be in the situation Berlin so much feared, in which positive liberty is gained by adjusting one's ambitions, desires, or aspirations.

Nonphysical force

Goodwin (2012) also opines that it is difficult to say that nudging is *coercive*, although, since it aims to exploit our cognitive weaknesses, it is surely *manipulative*. He calls it a subtle form of manipulation. I argue that once manipulation becomes effective enough it can be coercive. Few state explicitly that nudging is coercive, but Hausman and Welch (2010, p. 130) agree that it is "alarmingly intrusive" in that it diminishes our control over ourselves and our own evaluations of goals and alternatives.

The main question is whether or not nudging involves the use of force. For Goodwin (2012, p. 89), coercion involves an "attempt to pressurize a person (or persons) into adopting different behaviors, usually by force," and he argues that we cannot say that this definition applies to nudging. The reason is that he finds it difficult to find the use of force in a nudge. Once more, a more sophisticated understanding of power dispels such misperceptions (Sattarov, 2019). Even if his definition states that force is only usually applied, I will examine the use of nonphysical force, as this is the key element in play. We have already seen that Goodwin thought that negative liberty was only concerned with external physical obstacles. I disagreed there, and I argue that *psychological force* can be used to coerce. This is similar to what Faden and Beauchamp (1986, p. 355) call psychological manipulation, where "a person is influenced by causing changes in mental processes other than those involved in understanding."

Physical coercion is not particularly relevant in this context, but I argue in favor of broadening the concept of coercion to include phenomenon such as psychological coercion and other forms of manipulation related to efforts of behavioral modification. Hopper and Hidalgo (2006) write about psychological coercion, which is easily concealed and not easily understood. They argue that psychological coercion "can be as effective as physical violence in exerting control over a person" (Hopper & Hidalgo, 2006, p. 186).

We have already considered *threats*, for instance, in the example where you attempted to dissuade me by promising to shoot or shun me (Carter, 1999). I do not consider such attempts to influence behavior a threat to liberty when within legal bounds. However, nonnormative theories of liberty—those accompanied with a refusal to label some liberties more important than others, such as those of Carter (1999) and List and Valentini (2016)—would entail a different perspective on this issue. But what of the force of the better argument, then—if you use your force of logic to guide me toward a different goal? In such a process you would be appealing to my rational faculties and, should you succeed, it seems absurd to claim that my liberty is hurt. I cannot possibly claim freedom from arguments that change my opinion, even if there is power involved in all forms of persuasion (Sattarov, 2019).

But rhetoric is more than logic, so what if you mix some *pathos* into your persuasive endeavor? I might not perceive this, and I may change my actions, believing the cause to be your appeal to *logos*. No matter, as what you have done is both open and legitimate. People are more than logic, as rhetoricians have argued for nearly all of the written human history, and we have little grounds for saying that our *rational* faculties are more important than our *emotional* ones. Furthermore, persuasion is hardly ever *purely* rational, no matter how hard we try (Hausman & Welch, 2010). Antonio Damasio (2003, 2006, 2018) has written extensively on this topic. I follow him in his argument that reason and emotion are inseparable, and that the one makes little sense without the other. If we were to ban appeals to emotions, we would have to ban human communication and, all of a sudden, our position would be quite absurd.

This is an important point in relation to nudging, because nudging is based on the idea that our rationality is our authentic faculty. Any influence of emotions and other irrational mechanisms should be purged. I believe that this position is deeply problematic, partly because such a purge by nudging could take the form of manipulation, and partly because no one has the authority to decide that I should take more notice of some fictional rational process that is more present in me than my perceived feelings. You may attempt to rationally persuade me that you are right, but you cannot decide for me that *your* rationality is better than *my* desired mix of rationality, emotions, and various well or ill-founded influences.

"Best when invisible"

I have already touched upon the fact that nudges, according to Sunstein and Thaler, *should* be open to monitoring by the nudgees (Hausman & Welch, 2010). It is paradoxical, however, that a lot of the nudges work better when covert than when open and transparent (Goodwin, 2012). An obvious reason is that if people immediately recognized the goal of the nudge and agreed with it, rational persuasion would suffice, and there would be little reason to appeal to unconscious and irrational mechanisms in order to manipulate behavior.

Returning to the previous example: if, instead of creating associations between death, decay, and my goal, you said that these things were statistically correlated, I could hardly fault you. I would ask you why, and if you were able to convince me that these things were connected, you might succeed in diverting me. However, I would be free to consider your arguments null and void, and if so, no such associations and subconscious aversions to my goal would have been created.

Covert nudges must not be confused with rational persuasion, as they can be seen as an admission of the fact that rational persuasion is often ineffective. When someone openly targets the rational faculties of a person by providing them with facts and arguments, and attempts to persuade the person to act in a certain manner, this is persuasion, not nudging. If you can persuade someone in this manner, why use subterfuge to change their behavior?

Framing effects are one of the prime examples of nudging (Sunstein, 2016a). If you know that I have an irrational fear of snakes and that I love singing birds, you could portray the way toward my goal in two ways. You could either lament the scarcity of birds, and indicate that it is probably due to the existence of snakes in the area, which is factually true enough. Or you could alleviate my fear of snakes by saying that there is probably something like a 99.9% chance of me never seeing a snake here, and that the chances of meeting beautiful birds, just dying to sing to me, are just about as great. If you *told* me that you were framing these versions in different ways, and then provided me with the probabilities *for* and *against* whatever I liked or disliked, much of the effect would disappear. Framing is most effective when it is covert.

A nudge that is explained can be likened to rational persuasion, and is not of particular interest here. If we explain to a person that the choices have been rearranged, because of a public desire to reduce obesity and the fact the people in general have quite a sweet tooth, the rearranged menu would be just fine. Whenever we hide the nudge, it becomes a nudge proper, and is meaningfully separated from rational persuasion.

A final point to note is that nudging may be problematic due to people's individual differences (Goodwin, 2012; Mills, 2020b). What if I am a cognitive scientist with great knowledge of political science and nudging. My power to discover and evaluate the nudges I encounter would be quite substantial, whereas, in the case of someone who has never heard of either the disciplines or the nudge, the same power would probably be more limited. Goodwin (2012) points to nudging being unfair and, while I cannot pursue this topic here, it is possible to envisage a situation in which people with certain capacities are able to resist nudges and retain their original liberty, whereas, for others, the nudge truly becomes a coercive shove. I argue that everyone's ability to resist the nudge is threatened by Big Data and its application, but differences in such abilities will continue to exist until we are all unable to resist the nudge—a situation I do not consider particularly realistic. While interesting, my focus on the general effects opens for more detailed examination of the discriminatory and unfair results of Big Data-based nudging.

Nudging as (a) ineffective and (b) unavoidable, and (c) choice is exhausting

I will briefly consider three possible arguments against being concerned about nudging. The first is that it is ineffective, the second that it is unavoidable, and the third that choice is exhausting.

Tolerating nudging because it is ineffective

Hausman and Welch (2010) argue that nudging by nongovernmental agents is tolerable because it is ineffective, and also that it hard to regulate, but that is hardly a reason to resign philosophically. Furthermore, my argument is that nudging *is* effective, and is becoming more and more so, so this type of argument misses the mark on two accounts.

Nudging is subject to abuse, but Hausman and Welch (2010, p. 135) actually state that the main protection against this is "our limited proficiency at exploiting flaws in human decision making and the extent to which efforts at shaping choices on the part of different agents undercut one another." This reasoning fails for two reasons. Firstly, just because it is not effective does not make it acceptable, and secondly, we are becoming increasingly proficient at nudging, which is the main argument of this chapter.

Similarly, C. Mills (2013, p. 33) seems to argue that we should accept nudging because it is weak and too limited to be "genuinely transformative." While I argue that the problem with nudging is that it is a threat to liberty, C. Mills (2013, pp. 33–34) states that "the *true* failing of nudging in this regard is that it is often so benign that it will fail to be genuinely transforming because it cannot establish original (or significantly alter existing) moral or social norms."

Goodwin (2012) argues in a similar way, stating that a *problem* with nudging is that it does not stick and leaves people vulnerable to being nudged back. I posit that Big Data nudging is a form of secret coercion that is inimical to freedom because it *is* effective, not that it is not effective enough. Goodwin (2012, p. 90) has broader goals than the preservation of freedom, however, and for him, nudging "will not be enough." He argues for deliberative democracy, and as such I agree, not because nudging is not effective enough, but because rational persuasion and debate may be the only legitimate means of changing individuals' (lawful) actions.

While Goodwin (2012, p. 90) wants regulation of the "rampant commercialism and unregulated markets" that may make government nudges less effective, I argue for more regulation of both private *and* public nudging on behalf of both positive and negative liberty.

Tolerating nudging because it is unavoidable

One of the arguments Sunstein and Thaler use in favor of nudging is that it is *unavoidable* (Schmidt & Engelen, 2020; Sunstein, 2016a; Thaler & Sunstein, 2003). Hausman and Welch (2010) seem to agree with Sunstein and Thaler's proposition that manipulating choice architecture is less problematic than it might appear, since *some* organization of the choices we make is unavoidable. They explicitly state that "[w]hen choice shaping is not avoidable, it must be permissible" (Hausman & Welch, 2010, p. 132).

Amazon, for example, has to display *something* on its front page. Why, proponents of nudging ask, is choosing a front page that nudges us toward some specific products worse than any of the countless others' possible choice architectures? This is also related to the example in which you were blind. Let us, instead, now say that you had assigned me the task of preparing your outfit for the day, without further instructions. What should I now choose? Some choice is unavoidable, and following the logic established by Sunstein and Thaler and Hausman and Welch, I would now be free to choose what *I* considered to be in *your* best interest. If we dismiss the requirement that nudges have to be for the good of the nudgee, I would even be free to choose whatever I considered best—period. If I were in a naughty mood, and did not particularly like you, you could end up wearing a clown costume.

The obvious solution would be to elicit your preferences when I was given the assignment, so that I could choose what *you* wanted. In lieu of that, I could prepare two different outfits, and either tell you the difference. Me imposing my will on you is not made legitimate just by the fact that some choice has to be made.

I would even argue that me choosing an outfit at *random* would be less problematic in relation to your liberty than me deciding intentionally for you. Hausman and Welch (2010, p. 133) recognize the "important difference" between a set of alternatives intentionally designed to lead me in a certain direction and a random design. Random design would also lead me, but only the designed choice would involve the imposing "of the will of one agent on another," which is what I consider problematic as regards liberty (Hausman & Welch, 2010, p. 133). As I only consider the morally attributable actions of people potentially liberty reducing, this highlights the obvious and important differences between coincidental designs, designs by nature, etc.

Tolerating nudging because choosing is exhausting

I will only briefly consider the argument that nudging is good because *choosing* is supposedly too taxing. In the words of Sunstein (2016a, p. 61) "[i]n many areas, what the choice-making muscle needs is rest, not exercise." This was also proposed by Thaler and Sunstein (2003), and it is worthy of mention. Too many alternatives can be confusing, and a limited set of choices can *feel* liberating. While this can certainly be the case, it seems strange to argue that a constricted choice set increases liberty, and Berlin (2002) warned strongly about such approaches, which he identified with positive liberty. If that were the case, one would definitely have to adopt the positive conception of liberty, and see this as empowering.

Similarly, their focus on the default option drives them at times toward a stance that could lead them to being seen as champions of the freedom *not to choose*. It certainly seems paradoxical, but let us consider an example.

If I am selling you an online service, and I want you to choose a certain option that I know you would not have chosen without my interference, I could introduce many different choices in the hope that you would not be willing to sift through them all, and instead choose the default option that I want you to choose. This would of course be rather devious, but the chances are that I would be punished by you and other customers (Dholakia, 2016). It is, indeed, hard to regulate such tactics, but consumer protection laws, for example, can, and should, attempt to deal with them. Manipulative and misleading advertising, along with other kinds of objectionable manipulation by private companies or the government, should be regulated, even if it seems difficult to fully eradicate.

Similarly, when considering how to invest your savings for your old age, it would be quite taxing if you were required to consider and agree to each and every investment made by the fund in which you are investing. That is not a problem, however, as we very often consciously delegate choices to others. Such delegation is not nudging, and it is not problematic for liberty. When *you* decide that I should delegate, and remove my choices without me knowing, it becomes problematic.

Proposition 4.3

These considerations lead me to the third proposition about nudging and liberty.

> *Nudging is, in principle, inimical to liberty. Furthermore, the more effective it is, the more problematic it becomes.*

Nudging is problematic due to its covert and manipulative nature. A *strong* nudge becomes more of a coercive *shove*, and thus becomes even more problematic, particularly for proponents of negative liberty who are wary of interference and coercion. While it is easy to construct arguments against nudging based on positive liberty, it is also possible to construct arguments in favor of nudging based on the same concept. This leads me to consider negative liberty as the most interesting approach when analyzing the implications nudging has for liberty, but nondomination is also important in order to understand the need to safeguard such liberty.

When nudge comes to shove

The premises of my argument

I briefly summarize my propositions, which now become the three premises in my main argument:

- P1: Nudging is about influencing behavior, and it works
- P2: With Big Data, nudge comes to shove
- P3: Nudging is inimical to liberty, and more so as it becomes more effective

Big Data nudging as a threat to liberty

As noted in the discussion of premise 3, I do not consider positive liberty to be the most fruitful concept when considering how liberty is affected by Big Data nudging. While the focus on genuine autonomy might lead one to believe that nudging is definitely inimical to positive liberty, we must note that the concept of positive liberty opens the door to the kind of paternalism that Sunstein and Thaler espouse. If I am enslaved by my irrational passions, a well-constructed nudge that lets me be free from it might be seen as making me *freer*, even if it achieves its goals through the manipulation of unconscious processes instead of rational persuasion. Paradoxically, I can be *forced to be free* under this conception of liberty, which means that nudging and positive liberty have a less obvious relationship than one might at first imagine (Berlin, 2002, p. 179). While I consider the arguments against nudging based on loss of agency and not being one's own master as both strong and valid, other arguments in favor of paternalism and empowerment can, as shows easily, be constructed on the same foundations. I have discussed these issues above, and will thus proceed to the threats to liberty as seen through the lens of negative liberty and nondomination.

Based on the combination of premises 1, 2, and 3, I argue that strong and effective nudging constitutes a form of *coercion*. While not overt, it may constitute *secret* coercion, and coercion of any kind must be seen as inimical to liberty, and inimical to negative liberty as well. While liberals of all sorts are quite adamant about fighting overt coercion, I argue that it is time to take another look at the dangers posed by less open and obvious forms of interference. They might be a far greater threat to liberty than the obvious violations of rights that more readily grab the headlines. Such a position requires us to view manipulative efforts and the use of nonphysical force as constitutive of interference, to use Berlin's terminology.

John Stuart Mill is often portrayed as a utilitarian philosopher, but he is not a champion of the kind of utilitarianism that is engendered in the "libertarian paternalism" of Thaler and Sunstein (2003). While Mill found it permissible to stop a person from crossing a bridge that we (but not they) know is about to collapse, no person "is warranted in saying to another human creature of ripe years that he shall not do with his life for his own benefit what he chooses to do with it" (Mill, 1985, pp. 142,166). This is both because people are most valuable to themselves and others when free to develop novel ways of living and "experiments of living," and because, when it comes to their own lives, each person has "means of knowledge immeasurably surpassing those that can be possessed by anyone else" (Mill, 1985, pp. 120, 185).

However, while Hamburger (2001) clearly stated that Mill is opposed to interference and denial of choice, Mill himself opens up for "[c]onsiderations to aid his judgement, exhortations to strengthen his will," as long as the individual is, in the end, free to choose for himself (Mill, 1985, p. 143). Rational persuasion is fully compatible with liberty, since such efforts can "aid his judgement" and "strengthen his will." Nudging as contrasted with rational persuasion, on the other hand, is usually covert, and not as effective when made explicit. Mill opens up for appeals to a person's reason, not devices that play on cognitive "weaknesses" in order to achieve what we think a person really wants.

Conclusion

I acknowledge the existence of arguments in favor of nudging, but I conclude that it is problematic to use *liberty* as an argument for nudging. Such an argument would have to be based on the concept of positive liberty, empowerment, and emancipation from irrationality. However, even stronger arguments *against* nudging can be built on the same conception of liberty. Technology influences human decision-making in various ways, and it is important that we critically evaluate these effects (Dotson, 2012; Sattarov, 2019). Big Data-powered nudging has the potential to be both manipulative and coercive, and we should be wary of the effects such efforts have on liberty.

Most strands of liberal theory acknowledge the central findings from cognitive and behavioral theory that describe human beings' far from perfect capacity for rationality. The way we function makes us susceptible to manipulation and secret coercion, and a liberal theory can perfectly well demand regulation in order to prevent the exploitation of such weaknesses—and to protect individuals from abuse and coercion. Such an assumption does not imply that people are not autonomous or competent. It merely acknowledges the fact that we have various kinds and degrees of flaws and inclinations, and also that reason is not the only yardstick of authenticity. Liberal theory values liberty; in this entails a certain acceptance of unfortunate outcomes. In the spirit of Mill, we should to a certain degree be allowed to thrive or suffer from the results of the actions these inclinations lead us to, although the government has a role in making sure that it does not exploit our weaknesses unduly, and that other actors do not do so either.

While I do not accept arguments in favor of nudging based on liberty, it is easy to see that arguments based on utility could support nudging as utility will at times trump an absolute demand for liberty. However, I argue in favor of transparent traditional regulation and rational persuasion instead of nudging, when these approaches can serve the same purposes. Should we choose to nudge, we should not euphemize our efforts by claiming that we do so on behalf of freedom. Sunstein's (2016a) trap of abstraction is perhaps not as dangerous as the trap of consequentialist pragmatism. On a final note, the issue of regulation of nudging is beyond the scope of this book. I would argue, however, that we must not succumb to the temptation to state that, because it is difficult to regulate nudging, it is of little interest to discuss how to attempt to do so. Particularly since technology is making the problem at hand increasingly pressing.

References

Baruh, L., & Popescu, M. (2017). Big data analytics and the limits of privacy self-management. *New Media & Society, 19*(4), 579–596.

Bastiat, F. (1998). *The law*. Irvington-on-Hudson: Foundation for Economic Education.

Berlin, I. (2002). Two concepts of liberty. In H. Hardy (Ed.), *Liberty*. Oxford: Oxford University Press.

Calo, R. (2013). Digital market manipulation. *George Washington Law Review, 82*, 995.

Carter, I. (1999). *A measure of freedom*. Oxford: Oxford University Press.

Cohen, J. E. (2012). What privacy is for. *Harvard Law Review, 126*, 1904.

Coughlin, J. F. (March 27, 2017). *The 'Internet of Things' will take nudge theory too far*. Big Think.

Damasio, A. (2003). *Looking for Spinoza: Joy, sorrow, and the feeling brain*. Orlando: Harcourt Inc.

Damasio, A. (2006). *Descartes' error: Emotion, reason, and the human brain*. New York: Quill.

Damasio, A. (2018). *The strange order of things: Life, feeling, and the making of cultures*. New York: Pantheon Books.

Dholakia, U. M. (2016). *Why nudging your customers can backfire*. Retrieved from https://hbr.org/2016/04/why-nudging-your-customers-can-backfire.

Dotson, T. (2012). Technology, choice and the good life: Questioning technological liberalism. *Technology in Society, 34*(4), 326–336.

Eggers, W. D., Guszcza, J., & Greene, M. (July 26, 2017). *How government data can supercharge the nudge*. Governing.

Entis, L. (January 31, 2014). *The rise of 'nudge' advertising*. Entrepreneur. Retrieved from https://www.entrepreneur.com/article/231200.

Faden, R. R., & Beauchamp, T. L. (1986). *A history and theory of informed consent*. Oxford: Oxford University Press.

Giorgione, A. (September 21, 2017). *How big data is nudging us further than ever before*. Mumbrella.

Goodwin, T. (2012). Why we should reject 'nudge'. *Politics, 32*(2), 85–92.

Guszcza, J. (2015). The last-mile problem: How data science and behavioral science can work together. *Deloitte Review, 16*, 65–79.

Haggerty, K. D., & Ericson, R. V. (2000). The surveillant assemblage. *British Journal of Sociology, 51*(4), 605–622.

Hamburger, J. (2001). *John Stuart Mill on liberty and control*. Princeton University Press.

Hausman, D. M., & Welch, B. (2010). Debate: To nudge or not to nudge. *Journal of Political Philosophy, 18*(1), 123–136.

Helbing, D., Frey, B. S., Gigerenzer, G., Hafen, E., Hagner, M., Hofstetter, Y., ... Zwitter, A. (2019). Will democracy survive big data and artificial intelligence?. In *Towards digital enlightenment* (pp. 73–98). Cham: Springer.

Hobbes, T. (1946). *Leviathan*. London: Basil Blackwell.

Hopper, E., & Hidalgo, J. (2006). Invisible chains: Psychological coercion of human trafficking victims. *Intercultural Human Rights Law Review, 1*, 185.

Hugill, J. (2020). *Government nudging in the age of Big Data*.

Kittur, M. (2020). *Big data 'nudges' lead to better merchandise decisions*.

Kunstig intelligens. (2020). *Kunstig intelligens revolusjonerer casino-bransjen*.

Lepore, J. (2020). *If then: How the simulmatics corporation invented the future*. New York: W. W. Norten & Company.

List, C., & Valentini, L. (2016). Freedom as independence. *Ethics, 126*(4), 1043–1074. https://doi.org/10.1086/686006

Mill, J. S. (1985). *On liberty*. London: Penguin Books.

Mills, C. (2013). Why nudges matter: A reply to Goodwin. *Politics, 33*(1), 28–36.

Mills, S. (2020a). Nudge/sludge symmetry: On the relationship between nudge and sludge and the resulting ontological, normative and transparency implications. *Behavioural Public Policy*. https://doi.org/10.1017/bpp.2020.61

Mills, S. (2020b). Personalized nudging. In *Behavioural public policy*. https://doi.org/10.1017/bpp.2020.7

Oliver, A. (2013). From nudging to budging: Using behavioural economics to inform public sector policy. *Journal of Social Policy, 42*(4), 685–700. https://doi.org/10.1017/S0047279413000299

Oliver, A. (2015). Nudging, shoving, and budging: Behavioural economic-informed policy. *Public Administration, 93*(3), 700–714.

Packard, V. (1957). *The hidden persuaders*. New York: McKay.

Pettit, P. (1997). *Republicanism: A theory of freedom and government*. Clarendon Press.

Raz, J. (1986). *The morality of freedom*. Oxford: Clarendon Press.

Richards, N. M., & King, J. H. (2013). Three paradoxes of big data. *Stanford Law Review Online, 66*, 41.

Sattarov, F. (2019). *Power and technology: A philosophical and ethical analysis*. Rowman & Littlefield.

Schmidt, A. T., & Engelen, B. (2020). The ethics of nudging: An overview. *Philosophy Compass, 15*(4), e12658.

Sunstein, C. R. (2015). The ethics of nudging. *Yale Journal on Regulation, 32*, 413.

Sunstein, C. R. (2016a). *The ethics of influence: Government in the age of behavioral science*. Cambridge: Cambridge University Press.

Sunstein, C. R. (2016b). *Nudges that fail*. Available at: SSRN 2809658.

Thaler, R. H. (2018). Nudge, not sludge. *Science, 361*(6401), 431. https://doi.org/10.1126/science.aau9241

Thaler, R. H., & Sunstein, C. R. (2003). Libertarian paternalism. *American Economic Review, 93*(2), 175–179.

Thaler, R. H., & Sunstein, C. R. (2008). *Nudge: Improving decisions about health, wealth, and happiness*. New York: Yale University Press.

Watkins, J. W. N. (1967). *Hobbes's system of ideas: A study in the political significance of philosophical theories*. London: Hutchinson University Library.

Yeung, K. (2017). 'Hypernudge': Big Data as a mode of regulation by design. *Information, Communication & Society, 20*(1), 118–136.

5

The algorithmic tyranny of perceived opinion

Introduction

Never before have we had access to as much information as we do today, but how do we avail ourselves of it? In parallel with the increase in the amount of information, we have created means of curating and delivering it in sophisticated ways, through the technologies of algorithms, Big Data, and artificial intelligence. In this chapter, I examine how liberty is threatened by the way in which information is now handled.

I start by examining certain aspects of human psychology, such as the phenomenon called selective exposure, which denotes a human tendency to seek out information that supports preexisting beliefs, and to avoid unpleasant information that contradicts our opinions. I then discuss how information is curated in today's society and how digital technology has led to the creation of filter bubbles, while simultaneously creating closed online spaces in which people of similar opinions can congregate. While some liberals argue that more information can lead to exposure to new ideas and something akin to general enlightenment, I argue that we might be seeing the opposite happening in today's society. If people today, despite all the information that is available, are less exposed to ideas they consider undesirable, uncomfortable, and provocative, Big Data and algorithms might lead to less liberal societies, with individuals living in conditions that do not allow for the full development of individuality.

More specifically, I consider how certain effects created by Big Data, and the algorithms used to (a) deliver information to us and (b) deliver information from us to others, might exacerbate these problems. The issues discussed may not be very problematic if we regard liberty as only consisting in freedom from intentional physical interference. I, however, consider liberty to be a broader concept, and in this chapter I focus on liberty of opinion and the idea that liberty is as much a social as a political and legal phenomenon. With this understanding of liberty, the technological

Big Data's Threat to Liberty. https://doi.org/10.1016/B978-0-12-823806-6.00007-7
Copyright © 2021 Elsevier Inc. All rights reserved.

79

developments described above can be seen as posing a threat to individuality, autonomy, and the very foundation of liberal society. I focus in particular on the issues of perceived opinions and expectations and the renewed relevance of a Tocquevillian tyranny of opinion.

I mainly limit my discussion to the effects these developments have on individuals, but these issues are also of great importance to society in general, as there are clear connections to issues such as increasing polarization, the changing nature of capitalism, and the health of liberal democracy (Zuboff, 2019). Bozdag and van den Hoven (2015) provide an important account of how the phenomena I discuss relate to these broader societal issues, and to theories of democracy in particular. See also Sunstein (2018) for an analysis of the implications for deliberative democracy.

While Alexis De Tocqueville feared the tyranny of the majority, we would do well to fear the tyranny of the algorithms. In the modern age, we traverse a digital landscape that largely consists of opinions similar to our own, and in this sense the tyranny of opinion feared by Mill and Tocqueville may be more oppressive now than ever before.

Big Data and information

Big Data and artificial intelligence, Zuboff (2019) argues, are not necessarily problematic. She maintains that the true problem is the underlying system she refers to as surveillance capitalism. Large corporations that control the new technologies operate by a logic where there are few, or no, limits on the gathering and analysis of data for commercial purposes. It is how we employ technology, then, that is the main problem. A proper understanding of algorithms requires that we understand the "warm human and institutional choices" behind them, instead of seeing them as neutral technological phenomena (Gillespie, 2014).

It is important to be aware of the business models that form the basis for commercial applications of Big Data and algorithms. Foer (2017) provides an account of how the profit motive has changed the media landscape as we know it. The never-ending demand for more attention and more clicks is fed by deep insight from Big Data, which, in turn, is used to tune algorithms in ways that provide people with the kind of content that maximizes the variables the provider of information desires to be maximized. These variables are usually not our enlightenment, or the health of our democratic societies, and he fears that a "world without mind" will be the result (Foer, 2017). Bucher (2018) also describes how algorithms have changed news and journalism, while Gillespie (2010) shows how the rhetorical and discursive tactics of companies are often employed to give an air of neutrality. In the end, however, they are driven by the pursuit of profit. While Foer (2017) focuses on the monopolistic ambitions of the big tech companies, Zuboff (2019) argues that surveillance capitalism is something new that defies old analytical tools such as monopoly and privacy. These issues are of great importance, and we must understand these phenomena if we are to understand the true effects of technology.

In this section, I establish three propositions that will function as premises in my conclusion. They are related to (a) human nature and our tendency to seek information that corroborates, rather than challenges views we already hold, (b) algorithms and filter bubbles, and (c) the way in which we can now live digital lives in digital spaces inhabited to a large degree by like-minded people.

Human nature, selective exposure, and confirmation bias

Humanity is a peculiar bunch, and people who study our behavior tend to describe us as biased in our search for information. We are not fully rational, and there are many facets of our psychology that make us vulnerable to bad decision-making. These effects are discussed in detail in the literature on nudging (Sunstein, 2016; Thaler & Sunstein, 2003), and, for example, by Kahneman (2011). How these biases can be exploited using Big Data and AI was also discussed in detail in the previous chapter. Here, I focus more specifically on *selective exposure* and *confirmation bias*.

Selective exposure concerns individuals' propensity to seek out information that "aligns with their views and beliefs and avoid such content that is different in perspective or even challenging to their position" (Spohr, 2017, p. 153). This theory is from the 1960s, but it has received increasing attention as the algorithmic curation of information has become widespread, and is now "'one of the most commonly used theories in communication scholarship" (Garrett, 2009b, p. 677; Stroud, 2008).

The reason for its resurgence should be clear: personalization means that selective exposure has never been more relevant (Iyengar & Hahn, 2009). Explicit, or implicit, personalization makes the task of avoiding uncomfortable information, and obtaining information that supports our views, beliefs, and decisions, much easier than at any previous point in history

(Zuiderveen Borgesius et al., 2016). Personalization does not just make it *possible* to choose what one is exposed to; it *actually* leads to increasing selective exposure (Dylko et al., 2017). While there is disagreement about the strength of this effect, even people who tell us to relax say that the effect *is* there (Zuiderveen Borgesius et al., 2016).

A related phenomenon is cognitive dissonance, which is studied in the work of Festinger (1962). The reason we avoid information that conflicts with our beliefs is that it causes discomfort, and, as with other aversive stimuli and uncomfortable phenomena, we seek to reduce exposure to it (Fischer, Jonas, Frey, & Schulz-Hardt, 2005; Garrett, 2009b; Spohr, 2017). The work by Garrett (2009a, 2009b) suggests that the possibility of selecting what we are exposed to leads to greater exposure to supporting information, but not necessarily less exposure to information that conflicts with our existing beliefs.

As such, the theory of confirmation bias is proposed to be a better explanation of what actually happens in our new digital landscapes (Garrett, 2009b; Jonas, Schulz-Hardt, Frey, & Thelen, 2001). Confirmation bias explains the tendency we have to seek or interpret information in ways that support "existing beliefs, expectations, or a hypothesis in hand" (Nickerson, 1998, p. 175). This bias is not the result of conscious effort, and people are not aware that they are seeking and evaluating information in the way described here (Nickerson, 1998). Nickerson (1998, p. 175) describes how our unconscious selves at times work in a way resembling a lawyer building a case—not in order to shed equal light on all aspects of a case, but in order to *win*.

Iyengar and Hahn (2009) show that, in experiments, people prefer news sources they expect to agree with, both for political issues and "softer" issues, such as travel. They point to a growing body of evidence, and others, such as Jonas et al. (2001), have carried out experiments in which stronger selective exposure effects are found in settings that resemble real-life situations, with information being presented sequentially. Festinger originally considered selective exposure to be a problem related to conscious choice (Garrett, 2009b). With the rise of algorithms, *unconscious* selective exposure seems to be of even greater interest to us today.

Proposition 5.1

Individuals have a propensity to seek information that corresponds to their preexisting conceptions, through such mechanisms as selective exposure and confirmation bias.

Algorithms, curation of information, and "filter bubbles"

Attention is a scarce good, and even before the advent of online abundance of information, we all filtered, constantly (Kahneman, 2011; Sunstein, 2018). The main filtering tool is the algorithm, which is little more than computer code that transforms input into output (Rader & Gray, 2015). What makes algorithms interesting is that they are not, and cannot be, considered neutral (Dwork & Mulligan, 2013; Rader & Gray, 2015). The input could be *all* the news articles available, and the output could be what I see in my newsfeed in Facebook. Bucher (2012) provides a good account of how Facebook's algorithms choose what we see based on *affinity* (to authors), *weight* (of different kinds of material), and *time decay* (giving us more new stories). Of particular interest is her argument that these algorithms create clear rules for how to act in order to be successful (get a lot of attention), and quite severe sanctions for those who violate these rules (invisibility) (Bucher, 2012).[1]

While algorithms are automated and a machine technically chooses what I get to see, the choice is based on some initial instruction, and it is performed in pursuit of a set goal. As previously mentioned, I consider human beings responsible for the actions of machines and algorithms, no matter how complex and unpredictable they are (Sætra, 2021). Big Data is complex, and the way huge data sets is analyzed and used in conjunction with AI leads to a situation in which individuals subject to surveillance, analysis, and influence do not understand what occurs (Zwitter, 2014). Even worse, perhaps, is how the ethical issues involved are obscured—both unintentionally and intentionally—as those with decision power either are unaware or *attempt* to be unaware or the consequences of and responsibility for machine action (Sætra, 2021; Zwitter, 2014).

If I run a social media site and want to maximize my revenue, I will instruct my algorithm to provide you with as much pleasure as possible, in order to maximize the amount of time you spend on my site, which maximizes the revenue I receive, both in terms of advertising income and, most importantly, the data I collect about you. Like a map, or any other model of the world, an algorithm includes some things and leaves other things out, and what is left out is based on the "explicit and implicit values of their

---

[1]See also Bucher (2018) for a more complete account of the relationship between *algorithmic power* and politics, as well as a history of algorithms and an exposition on their technical aspects.

designers" (Dwork & Mulligan, 2013, p. 35). Gillespie (2010) describes how companies that employ algorithms have become "the primary keepers of the cultural discussion," and states that this is deeply troubling, due to their as yet quite unregulated operation. These digital "platforms" are not merely neutral areas in which information is freely given and acquired, since their owners seek the maximization of profit—they are not philanthropic well-doers (Gillespie, 2010).

What is interesting is that (a) people do not realize how these algorithms work and (b) this process might lead to what Pariser (2011) has called filter bubbles (Spohr, 2017). An algorithm works as a filter, and when this filter (a) only lets some information through, and (b) is used for most of an individual's information needs, the result is that the individual in question will be living, metaphorically, in a *bubble*. His perception of reality becomes skewed. Axel Bruns (2019b, 2019c) argues that filter bubbles as a concept is overrated, and that it has become a technological scapegoat that diverts attention from more serious phenomena. While I use *filter bubble* to explain the basic algorithmic operation described above, Bruns (2019a) defines a filter bubble as related to preferential communication, as a partner term to *echo chamber* (see below) which entails preferential *connection*. He also emphasizes the need to understand the psychological mechanisms that are seen as the more important problem, and I also focus heavily on such mechanisms in this chapter.

Customization and the prevalence of social networks and other sites that provide content through algorithms make this a real problem. The algorithms are driven by Big Data and the abundant information we all leave behind about our preferences and personalities. This means that sites such as Facebook can end up creating separate filter bubbles for many of us.

Here, I focus on two factors that make algorithmic curation of information problematic. Firstly, an algorithm, combined with Big Data and machine learning, may start out quite naïve, but as soon as we start making choices—liking some things, disliking others, spending a lot of time on specific kinds of content, etc.—the better the algorithms become at providing us with exactly what we desire. They thereby lead to a narrowing of the world we perceive. What we prefer is allowed through the filter, while things that upset us are filtered out. Secondly, the creators of algorithms program them in order to achieve certain goals. The specifics of these goals are undetectable to us and unrelated to our own preferences. This means that, even if a person did not exhibit selective exposure, the world they see through the algorithms by which they live will not be a neutral representation of the world.

The way algorithms shade the glasses through which we see the world is of increasing importance as customization and personalization increase (Zuiderveen Borgesius et al., 2016). Personalization can be *explicit*, in the sense that we actively help to train the algorithms by telling them to hide particular content, etc., or it can be *implicit*, as the algorithms simply learn from our behavior (Zuiderveen Borgesius et al., 2016). The effects of both kinds of personalization may be problematic, as the research of Rader and Gray (2015) shows. Social networks are at the forefront of the personal curation of information, and about 68% of Americans already get news via social media sites (Matsa & Shearer, 2018). One in five get news there often, and Facebook is by far the most popular network, as 43% of Americans get news via Facebook. Flaxman, Goel, and Rao (2016) note that the effects of social media are tempered by regular media. Zuiderveen Borgesius et al. (2016) have examined mainstream media outlets, and say that they are, as yet, still in the infancy of personalization. As such, they are not as likely to create algorithm-based filter bubbles.

Proposition 5.2

Algorithms powered by Big Data play an increasingly large part in curating the overwhelming amount of information that is available in modern society. Such algorithms are not neutral, and will always choose what information to present to users based on some predetermined logic.

Intergroup heterogeneity and intragroup homogeneity

The third factor I consider is that it is now possible for us to create digital spaces in which like-minded people can congregate. Historically, such contact has not been easy. When at work, with my family, or when shopping in the small town I live in, the chances of finding someone who shares my views on, and interest in, say, Thomas Hobbes, are not great. The fact that I constantly have to deal with people who do not agree with me could potentially be quite upsetting to me. I get little support for my opinions, and I constantly have to deal with everyone around me saying things that fit poorly with how I see the world. Previously, I might have attempted to relocate, say to a university with plenty of political theorists, or I might find a pen pal with whom I could share my views and interests. Either way, historically, I would be living most of my life interacting with real people with opinions very different from my own.

Today, however, I could simply create an online group for all the happy Hobbesians out there. I could also befriend these people, and after not too long I might have a social network that was predominantly composed of Hobbesians. The occasional family member and childhood friend might drop by of course, but they can easily be hidden from my newsfeed, making them merely hypothetical friends that I never see. It is important to note that, even if I do not explicitly hide these people, algorithms will be hiding them for me, often without me even realizing it, as a result, for example, of the algorithmic weight given to affinity as described by Bucher (2012).[2]

In this new digital space, I would see news stories shared by my fellows, their comments on these stories and other events, and their opinions on all sort of other things. This would *be* my new world. I might occasionally have to leave my house and briefly interact with real people, but I needn't place much emphasis on this, as my real life is now in my new digital world. Whenever I criticize a scholar for treating Hobbes as a straw man, I hear nothing but cheers, and when others post their Hobbesian analyses of current world affairs, the whole group applauds. We have created an *echo chamber* (Sunstein, 2018). And while my Hobbesian echo chamber might be considered quite benign, other echo chambers are far darker, and can consist of people who, for example, share a hatred of everyone who is different—in their opinions, politics, skin color, religion, or anything else.

An echo chamber is a space in which "individuals are largely exposed to conforming opinions" (Flaxman et al., 2016, p. 299). As such, we "inhabit different worlds" when we inhabit different echo chambers (Spohr, 2017, p. 152). In these worlds, we tend to share more of the information we believe to be conformable to group opinion, and so will the others (Flaxman et al., 2016). The consequence is increased polarization, as described by, for example, Sunstein (2018). Allcott and Gentzkow (2017) conducted a study showing that 20% of liberals' friends have the opposite ideological view, while the proportion for conservatives was 18%. One important aspect of digital media is that it is becoming easy to cater to much smaller groups (even individuals) than was possible with mass media. While mass media could cater to

---

[2]See Gillespie (2010, 2014) for a more thorough discussion of how nonneutral algorithms based on nonneutral datasets leave some things out, while highlighting others.

different positions in binary situations (blue/red, pro/con Brexit, etc.), the process described here divides these large groups into ever smaller groups, with great intragroup homogeneity. In large groups with little intragroup homogeneity, the effects described here will not be as severe, as there is a broad range of choices and opinions contained within them.

In echo chambers, people are comfortable, and while their own opinions are reinforced, they "lose the inclination to proactively discuss ideas with people or groups of a different position" (Spohr, 2017, p. 151). When we consider the fact that groups tend to take more extreme positions than individuals, this spells danger (Moscovici & Zavalloni, 1969; Spohr, 2017). We are less exposed to "cross-cutting content" on social media, and our *friends* are the most important factor: people choose like-minded friends who share news and opinions based on similar positions (Bakshy, Messing, & Adamic, 2015).

Consequently, we now belong to groups with a high degree of intergroup *heterogeneity* and a higher degree of intragroup *homogeneity*. When considered together with the two preceding phenomena, the compound effects we can envisage warrant a warning. Gerken (2004) calls this phenomenon *second-order diversity*, and Sunstein (2018) discusses how such a society can benefit from enclave discussions.[3]

Proposition 5.3

In the digital world, it is easy to create areas in which people can associate with like-minded people. In contrast to our lives in physical space, where (a) fewer like-minded people are accessible and (b) there are greater chances of encountering people with various beliefs, such areas can become what are referred to as echo chambers.

Freedom

Liberty and tyranny as political and social phenomena

There is no tyranny more dangerous than an invisible and benign tyranny, one in which subjects are complicit in their

---

[3]The issue of individual identity versus group identity falls outside the scope of this chapter, but I refer readers to Postmes, Spears, and Lea (1998) and Postmes, Spears, Lee, and Novak (2005) for a discussion of these issues.

victimization, and in which enslavement is a product of circumstance rather than intention.

Barber (1998, p. 582)

In order to evaluate the threats posed by the phenomena described in this chapter, I turn to Tocqueville and John Stuart Mill—two classical liberals who were concerned with liberty.

Tocqueville is perhaps most famous for his warnings against the effects of the tyranny of the majority. What struck him when visiting America was that freedom of discussion—and even freedom of thought itself—was limited, which led to a lack of "independence of mind" (Tocqueville, 2004, p. 293). This lack of independence of mind might be seen as a lack of liberty, particularly positive liberty. At the very least, it is a cause for concern about the *meaning* of both liberty in general and such terms as, for example, freedom of speech (Sunstein, 2018). Mill and Tocqueville both opine that people need access to a wide or full set of facts and opinions in order for these forms of freedom to be truly meaningful.

Tocqueville's concept of a tyranny of the majority is often thought to refer to the possibility that a small majority could subjugate and dominate the rest of the population. This is not the kind of tyranny I am concerned with, and I will instead speak of a *tyranny of opinion* when I discuss what I find to be most interesting in Tocqueville's concept.

When Tocqueville speaks of tyranny in America, he discusses tyranny as a social and cultural phenomenon (Horwitz, 1966; Maletz, 2002). Mill (1977, p. 178) reviewed Tocqueville's book, and he noted that Tocqueville's tyranny "is of another kind—a tyranny not over the body, but over the mind." Liberty does not just consist in political liberty and safety from a majority that creates poor laws or arbitrarily abuses physical power. It also consists in a liberty of *spirit* and *opinion*, and this liberty is threatened, or preserved, by society—not primarily through politics and the government. This form of liberty is particularly interesting with regard to the technologies I examine. Tocqueville describes America's "own unique brand of tyranny," where the tyrant is "the entire society itself, acting in concert without the need of oppressive laws" (Horwitz, 1966, p. 301). The lack of independence of mind is seen as the result of the "moral power" of the majority (Horwitz, 1966, p. 293; Tocqueville, 2004).

This moral power is not used to direct the actions of individuals through force, but to mold "their very natures" (Horwitz, 1966, p. 301). I argue that a tyranny "over hearts and minds" that manages to strip individuals of individuality cannot be considered less of a threat to liberty than a tyranny based on

Chapter 5 The algorithmic tyranny of perceived opinion **89**

physical oppression and interference in the more obvious sense (Horwitz, 1966, p. 302). As discussed in some detail in the previous chapter, I emphasize the need to take various sources of power into account (Sattarov, 2019).

Today we have a tyranny that has come to power right before our eyes—partly by superficial consent. Some might claim that the tyranny of opinion is something different from the facets of society Zuboff (2019) describes as surveillance capitalism. However, it makes sense to view, as Zuboff does, the rise to power of the surveillance capitalists as one of the main reasons for the developments leading to the increasingly social nature of all aspects of life, and the race toward a world where nothing is truly private. We could also invoke the concept of *societies of control*, in which individuals have become *dividuals*—divisible entities that can be recast in various forms (Deleuze, 1992).

However, even if such a tyranny is voluntary, it must be guarded against if liberty is to be preserved, particularly since such a tyranny is hard to perceive and may arise gradually without people noticing. Furthermore, as Zuboff (2019) notes, the initial phases of such a tyranny are marked by stealth and lawless conditions, and as we saw in Chapter 3, it is arguable that few people really give what would be considered informed consent when they agree to the terms required for access to what has become almost vital social infrastructure.

For Tocqueville, the very notion of *coercion* had to be redefined, as people did not recognize it in its classical form—the coercing tyrant was apparently nowhere to be found. Horwitz (1966) even uses the term psychic coercion to describe the way the majority forces everyone into something resembling uniformity. Coercion is usually thought to involve the use of physical compulsion, but, in lack of a better way to describe the psychological dimension of domination, I argue in favor of broadening the concept of coercion to also include such phenomena. This is in accordance with the discussion of the coercive potentialities of the combination of nudging and Big Data from the previous chapter. In America, *everyone* was the source of this tyranny—"each member of the community was both oppressor and oppressed"—and the instincts that make us rebel against injustice and oppression do not seem to be triggered by such a tyranny (Horwitz, 1966, p. 303). In Horwitz's words:

> *Coercion was not abhorrent because it was not really coercion;*
> *conformity was hardly onerous because it was freely chosen.*
> *Despotism, Tocqueville argued, had arrived at a new stage of*

> *perfection, since those who were oppressed glorified their oppression and honored their oppressor.*
>
> **Horwitz (1966, p. 303)**

Tocqueville himself says that he sought "in vain for an expression that exactly reproduces my idea of it and captures it fully ... [t]he old words '*despotism*' and '*tyranny*' will not do ... the thing is new" (Tocqueville, 2004, p. 818). Maletz (2002, p. 756) points to the penalty for opposing the tyranny in question, which consists of being "disregarded, ignored, overlooked"—in short, of being ostracized. "Chains and executions" are a thing of the past, Tocqueville says, as modern society "has today brought improvement to everything, even to despotism" (Tocqueville, 2004, p. 294). The modern tyrant does not punish unpopular utterings with death. Instead, he sentences dissenters to live as aliens in their own country, to the scorn of fellow citizens, and to being shunned "as one who is impure" (Tocqueville, 2004, p. 294). "Go in peace," the tyrant says, "I will not take your life, but the life I leave you with is worse than death" (Tocqueville, 2004, p. 294). In the modern world, some argue that this would be invisibility (Bucher, 2012).

Freedom requires more than me being allowed to swing my arms and legs around without restriction—it demands that I am also able to, so to speak, swing my tongue, and my *mind*. For me to be free, I must be able to say and think things that are contrary to popular opinion, without risking ostracism and heavy social sanctions. A tyranny that "ignores the body and goes straight for the soul" cannot be considered any less of a threat to liberty than its opposite (Tocqueville, 2004, p. 294). It would surely violate Pettit's conditions for nondomination (Pettit, 1997).

Proposition 5.4

> *Liberty requires, in addition to freedom from physical interference, an absence of social and psychological interference—domination—that blocks the freedom of spirit and opinion.*

Information and individuality

One of the main concerns of Mill in *On Liberty* (1985) was to convey the importance of being exposed to novel opinions, regardless of whether or not they contain truth, nontruth, or partial truth. He describes the "the quiet suppression of half" of the truth as the "the formidable evil," and writes that "there is always hope when people are forced to listen to both sides; it is when they attend only to one that errors harden into prejudices, and truth itself ceases to have the effect of truth by being exaggerated into falsehood" (Mill, 1985, p. 115).

For Mill, the idea of *individuality* is crucial, and individuality is important with regard to the developments I deal with here. Mill does, however, state that he regards "utility as the ultimate appeal on all ethical questions; but it must be utility in the largest sense, grounded on the permanent interests of man as a progressive being" (Mill, 1985, p. 70). I argue that utility also requires that we foster individuality, because in "proportion to the development of his individuality, each person becomes more valuable to himself, and is, therefore, capable of being more valuable to others" (Mill, 1985, p. 127).

One reason why *individuality* is important is that it prevents *subjectivity* and *communal values* from developing "in lockstep" (Cohen, 2012, p. 1911). Individuality is important to Mill, and Hamburger (2001, p. 4), when discussing Mill's ideas about *control*, states that "for Mill, interference, denial of choice, coercion, and encroachments on individuality are abhorrent." Individuality is both a precondition for and a guarantor of liberty in Mill's work.

Individuality requires that one has access to a variety of information—for inspiration, or simply to realize that something else is possible (Sunstein, 2018). Similarly, for it to be meaningful, liberty requires that there are options. If a society removes all options but one, but insists that people are free to choose whatever they desire, liberty has little meaning. A functioning liberal democracy also requires that citizens have access to a variety of information (Barber, 1998). In order to make reasonable choices, awareness of a broad set of "opinions and options" is required (Bozdag & van den Hoven, 2015). These concepts are related, in that one could argue that, if liberty is to be preserved, one has to provide the individual with the opportunity to (a) be informed about the world in which he lives and (b) to be free to choose the way in which he wants to live. Considering our tendency towards selective exposure, and how companies program their algorithms to maximize profit, we might even argue that, in addition to access to a broad array of information, one should also seek to attain a certain level of minimum exposure to this information.[4]

---

[4]Sunstein (2018) enters a plea for serendipity and examines how we can counter the effects I discuss. See also Bozdag and van den Hoven (2015) for a discussion of some ways to technically combat filtering effects.

Proposition 5.5

Access and exposure to a broad array of information is a requirement for the development of individuality.

The threat of a tyranny of perceived opinion

Turning the propositions into premises

I have thus far put forth five propositions that will now be considered as premises for assessing the threat here analyzed. A summary of the premises is presented here, but please refer to their full statement for a more precise rendition of the complete argument.

- Premise 5.1 (P1): Individuals have a propensity to seek information that supports their preexisting conceptions (and to avoid information that conflicts with the same).
- Premise 5.2 (P2): Algorithms, which are not *neutral*, function as curators of information in modern society.
- Premise 5.3 (P3): Digital technology and social networks allow people with similar opinions to congregate, and tend to make them do so, minimizing the need to deal with people with conflicting opinions.
- Premise 5.4 (P4): Liberty requires, in addition to freedom from physical interference, an absence of social and psychological interference that blocks freedom of spirit and opinion.
- Premise 5.5 (P5): Access and exposure to a broad array of information is a requirement for the development of individuality.

My first conclusion is that premises 1, 2, and 3, when combined, create great potential for living lives in which one is predominantly exposed to information that supports preexisting opinions and minimizes contact with opposing views. Algorithms are not neutral and may distort and restrict the information we receive, and tend to do so, both unintentionally and for the purpose of achieving some goal of their creator. When our views are further driven to narrowness by the human tendency to prefer information that confirms our preconceptions, our window to the world becomes a slit. When, in addition to this, individuals also congregate with like-minded people and get *social* support for their opinions, the potential for citizens to live with radically different perceptions of reality becomes quite great. The difference, then, is apparently *intergroup*, while there is great *intragroup* homogeneity. I refer to this as the *compound confirmation effect* of information technology and social networks. When coupled with premise 4, the threat to liberty is clear, and this is also pursued in more detail in the concluding chapters.

The second conclusion I wish to draw is that the effects of premises 1, 2, and 3, combined with premise 5, point to a threat to individuality. If the compound confirmation effect just described makes it difficult to have access to varied information, individuality is threatened. And, if individuality is lost, liberty makes little sense. The main conclusion is that, when all the premises are considered, they paint a picture of a substantial threat to liberty.

The tyranny of perceived opinion

If the premises established above are accepted, the threat to liberty emerges as a two-headed creature. The two heads are somewhat similar, but the creature in question is a hybrid, with one human head and one that is the head of a machine.

The human side of the threat is nothing short of a resurrection, with a vengeance, of the phenomenon described by Tocqueville. His tyranny is most prevalent when the moral power of popular opinion is (actually, or is perceived to be) strong, and it seems evident that this power becomes greater as groups become more and more homogenous. The cumulative force of uniformity is heightened as more and more people get in line.

It may seem paradoxical to claim that modern society is homogenous, and this is where I argue that we have a *new* form of Tocquevillian tyranny that follows from the increased within-group homogeneity that follows from the tendency of like-minded people to gather in what have been referred to as echo chambers.

What happens on Facebook when one of your friends posts a rant that greatly displeases you? You immediately feel your heart rate increasing and your hands becoming moist. An opinion radically different from yours has triggered something in you, and you have three options. You can (a) ignore it, (b) argue with the person, or (c) hide the person. If you become sufficiently annoyed, alternative (a) is hard, and you know that this person will most likely post something similar again quite soon. Alternative (b) is an often-advocated option, but it is a costly option. You will have to engage with someone very different from yourself, and you will be at risk of losing a very public debate in front of all of your, and his, friends. Option (c) thus becomes an attractive option. If I simply tell Facebook that I do not want to see this person anymore, I want to *hide* this person from my sight, the problem disappears.[5]

But how does knowing all this affect how I act? If many people choose alternative (c), is the result really that different from the punishment Tocqueville's tyrant meted out? The dissenter was shunned and ignored—put out of sight, and not even dignified with a response. The possibility of such punishment makes most of us wary of what we say and proclaim (Bucher, 2012). The moral power here involved is related to the psychological force discussed in the previous chapter, and any type of liberty that is unable to account for the exertion of such force is insufficient. Showing how a theory of liberty can account for this is the topic of Chapter 7.

In addition, as human beings we have a strong desire to be, or at least appear to be, consistent (Nickerson, 1998, p. 197). Every time I choose not to state something I believe to be controversial, or every time I half-heartedly "like" something just because I realize that it is expected of me, I may commit myself to the mainstream position. If so, the problem of selective exposure will be increasingly severe, and I will, unconsciously, have locked myself into the cage of the opinions of my group (Moscovici & Zavalloni, 1969). Bucher's (2012) discussion of how sites such as Facebook can create a fear of invisibility is highly relevant in this respect, and so is the fact that while we may not actively tell Facebook to ostracize those we do not like, the mere fact of indicating little affinity will lead the algorithms to do it for us.[6]

I now turn to the mechanical head of my two-headed creature. The problem, in short, is that algorithms and Big Data can be seen as partly determining our development, while also removing the opportunity to see alternatives to, and resist, the path with which we are presented (Baruh & Popescu, 2017). Baruh and Popescu (2017) are concerned with the power imbalances created between individuals and institutions, and they

---

[5]I must note that I do not consider the more active ways in which a tyranny of opinion can be amplified in the digital sphere. Blackford (2018) provides a useful and recent discussion of issues such as online moral police, online shaming, and what he calls cybermobs. While Tocqueville's tyrant can employ such tactics, he may not need to when the perception of prevailing opinion and expectations are strong.

[6]The details of how users change their behavior in order to convince the algorithms to make them visible are described by Bucher (2018) and Turkle (2017).

fear that the dynamics involved in employing Big Data may shape and influence individuals by creating self-fulfilling prophecies. When Big Data leads companies to predict preferences and actions, these very predictions may lead people to live up to them (Baruh & Popescu, 2017). For example, a search provider knows much more about us than the query we enter at a certain point in time, and uses these data to predict and anticipate our desires (Gillespie, 2014). Using knowledge about us to improve a specific service is behavioral feedback, but the data are used for much more, creating what Zuboff refers to as behavioral surplus (Zuboff, 2019). This leads to problems deciding whether an algorithm is successful because it provides the correct output, or because the users adapt to the output, as discussed by Lanier (2014).

It is important to keep in mind that algorithms can be used to *intentionally* influence actors as well, a topic I examine in more detail in the following chapters. Dwork and Mulligan (2013, p. 38) use the language of freedom, as they state that the process of Big Data and the subsequent classification shackle users to profiles that are then used to determine what is relevant for them, a process that relates to Deleuze's (1992) concept of the dividual. The algorithms are "prediction engines, constantly creating and refining a theory of who you are and what you'll do and want next" (Pariser, 2011, p. 10). That such profiles are not accurate representations of a person seems obvious, since some information is gathered, some is approximated, and a lot is considered unimportant for the purposes of the profile manager (Gillespie, 2014). My argument is that such predictions, and the expectations they create, are coupled with the tyranny just discussed, and I focus on three ways in which we might go from the seemingly useful to something more ominous.

Conclusion

Alexis De Tocqueville feared an immaterial tyranny of public opinion—a tyrant who was hard to perceive, and whose tyranny was a seemingly benign and accepted reign. He showed that such a tyranny was no less of a danger to liberty than the tyranny of old. I argue that a tyranny of minds, not bodies, that people consent to because they do not perceive or understand what is happening, is an even greater danger than a tyranny that we would all immediately recognize as such. In this chapter, I show that we would be wise to heed the warnings of Tocqueville in our own day and age, even if the tyranny he described now appears in a slightly different form.

The threat of tyranny, which is a threat to liberty, has arisen because (a) we tend to seek facts that correspond to our preexisting opinions, (b) we use algorithms as curators of information, and (c) it is now easy for us to create networks in which we mainly, or exclusively, associate with people like us. The compound exposure effect created by these three factors is what makes Tocqueville's warning prescient. In our echo chambers and filter bubbles, the tyranny of perceived opinion is a clear and present danger to liberty.

While negative liberty is perhaps not most threatened, positive liberty requires a form of autonomy that can easily require that we have access to the information we need to be autonomous citizens. Pettit's (1997) concept of nondomination is very much applicable to our algorithms, as they, and their creators, clearly have the capacity to interfere arbitrarily in the choices we make when we seek to understand the world and make the choices we do in our lives (Bozdag & van den Hoven, 2015). I have deliberately chosen not to give a central place to discussions of the possibility of abusing algorithms in this chapter, as I believe the threat to liberty is clear enough even without this form of manipulation and coercion. However, when algorithms are weaponized and used in conjunction with surveillance and behavioral modification, we see the three threats of this book coming together. That is the topic of the next chapter.

In addition to the threat to liberty, we should also take note of the fact that a liberal and healthy society seems to require well-informed citizens with a similar enough worldview to enable them to discuss, debate, and deliberate together in order to find a direction ahead for society. This is a concern even if we are not inclined to view deliberative democracy as the main concern (Sunstein, 2018). The decreased breadth of the information we receive due to the situation I have described here has clear implications for the development of each individual's individuality and liberty, and I have restricted myself to this level. The threat to society posed by these developments is of great importance, and it is paramount that they are given more attention in the disciplines of political science and law, so that they do not remain in the exclusive domain of communication studies and information technology.

References

Allcott, H., & Gentzkow, M. (2017). Social media and fake news in the 2016 election. *The Journal of Economic Perspectives, 31*(2), 211–236.

Bakshy, E., Messing, S., & Adamic, L. A. (2015). Exposure to ideologically diverse news and opinion on Facebook. *Science, 348*(6239), 1130–1132.

Barber, B. R. (1998). Three scenarios for the future of technology and strong democracy. *Political Science Quarterly, 113*(4), 573–589.

Baruh, L., & Popescu, M. (2017). Big data analytics and the limits of privacy self-management. *New Media & Society, 19*(4), 579–596.

Blackford, R. (2018). *The tyranny of opinion: Conformity and the future of liberalism.* Bloomsbury Publishing.

Bozdag, E., & van den Hoven, J. (2015). Breaking the filter bubble: Democracy and design. *Ethics and Information Technology, 17*(4), 249–265.

Bruns, A. (2019a). *Are filter bubbles real?* John Wiley & Sons.

Bruns, A. (2019b). Filter bubble. *Internet Policy Review, 8*(4).

Bruns, A. (2019c). *It's not the technology, stupid: How the 'Echo Chamber' and 'Filter Bubble' metaphors have failed us.*

Bucher, T. (2012). Want to be on the top? Algorithmic power and the threat of invisibility on Facebook. *New Media & Society, 14*(7), 1164–1180.

Bucher, T. (2018). *If... then: Algorithmic power and politics.* Oxford University Press.

Cohen, J. E. (2012). What privacy is for. *Harvard Law Review, 126*, 1904.

Deleuze, G. (1992). Postscript on the societies of control. *October, 59*, 3–7.

Dwork, C., & Mulligan, D. K. (2013). It's not privacy, and it's not fair. *Stanford Law Review Online, 66*, 35.

Dylko, I, Dolgov, I, Hoffman, W., Eckhart, N., Molina, M., & Aaziz, O. (2017). The dark side of technology: An experimental investigation of the influence of customizability technology on online political selective exposure. *Computers in Human Behavior, 73*, 181–190. https://doi.org/10.1016/j.chb.2017.03.031

Festinger, L. (1962). *A theory of cognitive dissonance.* Stanford: Stanford University Press.

Fischer, P., Jonas, E., Frey, D., & Schulz-Hardt, S. (2005). Selective exposure to information: The impact of information limits. *European Journal of Social Psychology, 35*(4), 469–492.

Flaxman, S., Goel, S., & Rao, J. M. (2016). Filter bubbles, echo chambers, and online news consumption. *Public Opinion Quarterly, 80*(S1), 298–320.

Foer, F. (2017). *World without mind.* Random House.

Garrett, R. K. (2009a). Echo chambers online?: Politically motivated selective exposure among internet news users. *Journal of Computer-Mediated Communication, 14*(2), 265–285.

Garrett, R. K. (2009b). Politically motivated reinforcement seeking: Reframing the selective exposure debate. *Journal of Communication, 59*(4), 676–699.

Gerken, H. K. (2004). Second-order diversity. *Harvard Law Review, 118*, 1099.

Gillespie, T. (2010). The politics of 'platforms'. *New Media & Society, 12*(3), 347–364.

Gillespie, T. (2014). The relevance of algorithms. In T. Gillespie, P. Boczkowski, & K. Foot (Eds.), *Media technologies: Essays on communication, materiality, and society.* Cambridge: MIT Press.

Hamburger, J. (2001). *John Stuart Mill on liberty and control.* Princeton University Press.

Horwitz, M. J. (1966). Tocqueville and the tyranny of the majority. *The Review of Politics, 28*(3), 293–307.

Iyengar, S., & Hahn, K. S. (2009). Red media, blue media: Evidence of ideological selectivity in media use. *Journal of Communication, 59*(1), 19–39.

Jonas, E., Schulz-Hardt, S., Frey, D., & Thelen, N. (2001). Confirmation bias in sequential information search after preliminary decisions: An expansion of dissonance theoretical research on selective exposure to information. *Journal of Personality and Social Psychology, 80*(4), 557.

Kahneman, D. (2011). *Thinking, fast and slow.* Macmillan.

Lanier, J. (2014). *Who owns the future?* Simon and Schuster.

Maletz, D. J. (2002). Tocqueville's tyranny of the majority reconsidered. *The Journal of Politics, 64*(3), 741–763.

Matsa, K. E., & Shearer, E. (2018). *News use across social media platforms 2018* (p. 10). Pew Research Center.

Mill, J. S. (1977). *Essays on politics and society* (Vol. 18). Toronto: University of Toronto Press.

Mill, J. S. (1985). *On liberty.* London: Penguin Books.

Moscovici, S., & Zavalloni, M. (1969). The group as a polarizer of attitudes. *Journal of Personality and Social Psychology, 12*(2), 125.

Nickerson, R. S. (1998). Confirmation bias: A ubiquitous phenomenon in many guises. *Review of General Psychology, 2*(2), 175–220.

Pariser, E. (2011). *The filter bubble: What the internet is hiding from you.* UK: Penguin.

Pettit, P. (1997). *Republicanism: A theory of freedom and government.* Clarendon Press.

Postmes, T., Spears, R., & Lea, M. (1998). Breaching or building social boundaries? SIDE-effects of computer-mediated communication. *Communication Research, 25*(6), 689–715.

Postmes, T., Spears, R., Lee, A. T., & Novak, R. J. (2005). Individuality and social influence in groups: Inductive and deductive routes to group identity. *Journal of Personality and Social Psychology, 89*(5), 747.

Rader, E., & Gray, R. (2015). *Understanding user beliefs about algorithmic curation in the Facebook news feed.* Paper presented at the proceedings of the 33rd annual ACM conference on human factors in computing systems.

Sætra, H. S. (2021). Confounding complexity of machine action: A Hobbesian account of machine responsibility. *International Journal of Technoethics, 12*(1).

Sattarov, F. (2019). *Power and technology: A philosophical and ethical analysis.* Rowman & Littlefield.

Spohr, D. (2017). Fake news and ideological polarization: Filter bubbles and selective exposure on social media. *Business Information Review, 34*(3), 150–160.

Stroud, N. J. (2008). Media use and political predispositions: Revisiting the concept of selective exposure. *Political Behavior, 30*(3), 341–366.

Sunstein, C. R. (2016). *The ethics of influence: Government in the age of behavioral science.* Cambridge: Cambridge University Press.

Sunstein, C. R. (2018). *# Republic: Divided democracy in the age of social media.* Princeton University Press.

Thaler, R. H., & Sunstein, C. R. (2003). Libertarian paternalism. *The American Economic Review, 93*(2), 175–179.

Tocqueville, A. D. (2004). *Democracy in America.* New York: The Library of America.

Turkle, S. (2017). *Alone together: Why we expect more from technology and less from each other.* UK: Hachette.

Zuboff, S. (2019). *The age of surveillance capitalism: The fight for a human future at the new frontier of power: Barack Obama's books of 2019.* New York: PublicAffairs.

Zuiderveen Borgesius, F., Trilling, D., Möller, J., Bodó, B., De Vreese, C. H., & Helberger, N. (2016). Should we worry about filter bubbles? *Internet Policy Review. Journal on Internet Regulation, 5*(1).

Zwitter, A. (2014). Big data ethics. *Big Data & Society, 1*(2). https://doi.org/10.1177/2F2053951714559253.

6

The three threats in concert

Introduction

Big Data's threat to liberty is threefold, and in the preceding chapters I have presented and examined these related, but clearly different, threats. While I have shown that the threats are serious enough in isolation, the negative potential of Big Data emerges more clearly when they are seen in concert. It is time to examine in more detail how they relate to each other, and also how a proper understanding of the nature of privacy might enable us to more effectively face them all.

The first threat discussed was surveillance. Big Data enables surveillance, which is a threat both due to our right to privacy, and because surveillance itself constitutes a liberty-reducing interference in our lives. This threat to liberty highlights the importance of privacy and its relation to the use of surveillance for gathering data. The data is subsequently used in ways that exacerbate the threat, *and* it creates the foundation for the other threats.

The second threat is how Big Data is used to build highly detailed personality profiles, which in turn are used to influence us in various ways through the techniques of nudging. These nudges might be used for ill or good purposes. Regardless of motive, manipulation is involved, and I have argued that such nudging can even become coercive. Through various forms of nudging, surveillance, which is itself a threat to liberty, enables a form of coercion which is a separate, but related, threat. When employing Big Data nudges, we use algorithms both to determine what sort of nudge is most effective for each person, *and* how these nudges should be delivered to each person. The threats described in Chapter 4 relate to the combination of surveillance and nudging.

The third threat emerges as the data gathered through monitoring our every move is also used to determine what sort of information we receive from the world around us. When data about us, both as individuals and groups, are fed through algorithms, we run the risk of supercharging the human tendency for seeking confirmatory information, which can lead to the creation of filter bubbles. A similar tendency in social networks leads

Big Data's Threat to Liberty. https://doi.org/10.1016/B978-0-12-823806-6.00004-1
Copyright © 2021 Elsevier Inc. All rights reserved.

to the creation of echo chambers, and suddenly the Big Data-driven curation of our digital landscapes has led to both a narrowing of the information we receive and a reduced ideological diversity among the people we interact with. Surveillance enables the collection and subsequent employment of information in these ways, and Chapter 5 thus already deals with the combination of surveillance and algorithmic curation of data.

Of most interest here, then, is the nature of the relationship between nudging and the algorithmic curation of data. One challenge and threat that was not explored in Chapter 5 was how the use of algorithmic curation of information may be "weaponized," as the predictions created through the process described can be used to intentionally modify an individual's behavior. Algorithmic curation of information and the design and presentation of personality profiles can thus become an instrument for nudging. The potential impact of these phenomena is, as shown, further increased by surveillance. In concert, the three threats lead to societies in which we lose liberty through surveillance and lack of privacy, lack of alternative lifestyles, manipulation and coercion through the exploitation of the information gathered by surveillance, and a reduction of the breadth of information that allows us to develop into autonomous individuals.

In addition, seeing the compound threat helps highlight the key role played by privacy, as privacy is instrumental in effectively facing all three threats. While regulatory approaches to Big Data and privacy are not the focus of this book, the latter part of this chapter presents the argument that privacy is a public good, and that a proper understanding of this fact will have clear consequences for the regulation of Big Data and privacy and for the restatement of liberty and in the next chapter.

Nudging by curation of information

One might view the techniques of nudging and the mechanisms behind filter bubbles as two facets of the same phenomenon: both are made possible by surveillance and the creation of in-depth personality profiles. While nudging is about the intentional use of such information in order to influence our actions, filter bubbles and echo chambers can be seen as the potentially unintended consequences of having algorithms provide us with what someone believes we will like. When the information and the suggestions we receive and the way we are portrayed creates expectations. They become *leading*, and thus of interest in relation to liberty. The theories of expectations here presented are useful additions to the current debate on how our use of Big Data and algorithms is nonneutral, manipulative, and shaping. While biased expectancies can be formed without any underlying ill or good intentions, I here focus on how these expectancies can be actively used to shape an individual's perception and understanding of themselves and the world around them. This, in turn, allows for the shaping of their behavior.

Three theories from psychology and sociology help elucidate hitherto less explored facets of the threat Big Data poses. It relates to the effects Big Data, when combined with nudging and algorithmic curation of information, has on the free development of individuals. The theories are (a) the expectancy effect, (b) the Proteus effect, and (c) the Thomas theorem.

The expectancy effect

The expectancy effect, or behavioral confirmation theory, describes a phenomenon whereby my expectations of someone actually influences how he or she acts and performs (Rosenthal, 1994; Rosenthal & Rubin, 1978; Snyder & Swann, 1978). It comes into play as soon as individuals experience individualized profiles on websites, entertainment services, etc. Data is used to create an environment that the service provider expects us to be comfortable with, and this involves everything from the subtle organization of content, the visual presentation of the content, and the various forms of suggestions offered. The entire service is experienced as custom-made for each individual, which makes their expectations more likely to have the effects described as expectancy effects or behavioral confirmation. People may perceive the suggestions and selections they are offered on sites such as Facebook and Netflix as being based on relatively objective criteria, such as their usage history, but that is only partly the case. They are based on history, what people similar to the user in question tend to do, and on what the provider wants the user to consume (Gillespie, 2014). Once again, we must keep the nonneutrality of algorithms in mind.

This creates what is perceived as an expectancy, which can become self-fulfilling. It also constitutes a fundamental challenge to evaluating the effectiveness of algorithmic prediction, as the predictions have effects on the outcomes they are intended to predict. We usually evaluate the accuracy of our predictions by seeing how well actual behavior conforms to our predictions (Baruh & Popescu, 2017; Lanier, 2014). If the predictions themselves *cause* behavior to change, this is clearly not a very useful measure of accuracy.

An obvious way to nudge by expectations is to present an individual with a prediction of what they are about to do, while implicitly or explicitly relaying the message that the prediction is based on knowledge of the person and what we expect of them. On a shopping site, for example, this is done implicitly by tailoring the site to each individual and showing them what they have previously done and what others like them tend to do on the site. As soon as a medium is experienced as personalized, there is an implicit message that the personalization is based on knowledge of the user. This creates a setting in which the user is prone to perceive suggestions as expectations, and, for example, suggestions for new books to read as an expectation that they—or someone like them—*should* read these books. This allows for subtle nudges that intimate "honest" predictions not mixed with an intent to change behavior. The best effects are likely achieved when the nudges are only slightly shifted and altered from what the machine learning models would have output if maximum fidelity to the target was desired, as the user will then most identify with the expectations provided.

The Proteus effect

The Proteus effect describes how people might change their behavior as a consequence of how they perceive themselves. Changing their appearance in a digital setting also changes their behavior (regardless of how, or whether, others see them) (Yee & Bailenson, 2007). The Proteus effect, which is based on the self-perception theory of Bem (1972), suggests that "people often infer their own attitudes and beliefs by observing their behaviors in the same manner they would observe another person" (Fox, Bailenson, & Tricase, 2013, p. 932).

This effect describes behavioral change resulting from changes in how a person is portrayed in a virtual environment, for example, as an avatar—a digital representation of the self—in a computer game or an online casino (Baruh & Popescu, 2017). When the avatar changes, the behavior of the user also changes. For example, when I am portrayed as attractive, I tend to move closer to others and disclose more personal information, and when made tall I tend to be more confident in negotiations (Yee & Bailenson, 2007). Furthermore, when women are portrayed in a sexualized manner, they become more prone to self-objectification, and even report higher acceptance of rape myths (Fox et al., 2013).

How then is this related to Big Data? Changing the online representations of the self—arguably one of the most important ways to portray oneself in our age—can also change behavior and feelings of self. In social media, people are portrayed in various ways, and they are also made aware of what is shown to, and liked by, other people. The framework for this representation of self is controlled by companies such as Facebook, and what they choose to highlight to my friends from my posts is also controlled by them. Bucher's (2012) fear of invisibility may be the result of a conscious effort by companies to promote a certain user behavior, and by portraying the user in different ways they may effect real changes in user behavior, and not just behavior, as a change in how a person is portrayed might change their very perceptions of themselves. As with the expectancy effect, it seems reasonable to assume that *slight* shifts in representation have the most potential for changing perceptions of the self, while more obvious changes to characteristics not intimately tied to a person's perceived identity might still produce shifts in behavior. Another and more straightforward example of how to nudge in order to change behavior would be for a bank to implement a visual representation of the bank's client on their user profile pages, and subtly changing aspects like the character's clothing, physical features, and general appearance in order to encourage spending, for example. When this is combined with knowledge of the person in question, it enables the creation of characters that are both believable and able to subtly portray and play on the various hopes, goals, and aspirations of the client as far as this is known or inferred.

The Thomas theorem

The final theory to keep in mind is the Thomas theorem, which in short-form details self-fulfilling prophecies (Merton, 1948). The theory states: if people define situations as real, the consequences of those situations *are* real. If I portray a situation—a prediction or prophecy—and get you to believe in it, there is a greater chance of my prediction actually coming true than there would otherwise have been. Merton (1948) gives examples such as banking crises happening because of rumors about banks having problems or a student failing a test because they worry and believe they will fail. Merton points to the ethnic conflicts where those referred to as the "whites" fail to see that "he and his kind have produced the 'facts' which he observes"—the facts that prolong the conflict (Merton, 1948, p. 196).

When algorithms present us with clear expectancies, we may perceive these expectations as prophecies that we, in turn, make reality. When a shopping suggestion is presented in such a manner that it is perceived to be a description of a future fact, we move from the previously discussed expectations to the realm of more clearly self-fulfilling prophecies. Furthermore, as Big Data is based on a form of "radical empiricism" (Pearl, 2021), predictions will have a conservative force that promulgates old definitions of how things *are*, despite the possibility that things could very well have been different (Sætra, 2018).[1]

The Thomas theorem shows how algorithmic prediction could lead to *external* impediments to individuals' liberty, but, in combination with the two other theories, it seems clear that such prediction can simultaneously have internal and external effects. Assuming that these effects are real, it seems uncontroversial to assume that the level of personalization of expectations and representations made possible by Big Data and algorithms makes these effects stronger.

While honest attempts to describe a situation might lead to situations in which these attempts cause the situation, it can also be used intentionally to influence actions and increase the likelihood of certain outcomes. Those in control of the algorithms might create situations in which other individuals are led to fulfill their prophecies. During a conflict, for example, people might be led to believe that others like them are just now about to gather in front of congress to arrange a demonstration against the unjust usurpation of power. When combined with cues derived from personal data, this creates the expectation that the individual seeing this will be part of the demonstration and the prophecy will be more effective in terms of self-fulfillment. This technique might also be used to encourage certain shopping, or even voting, behavior. One simple way to make a prediction seem like a matter of fact is to insert cues that relate to the situation after the desired action has already been taken. For example, suggestions for a nice diner to visit after vote has been cast mixed with cues regarding how the individual will purportedly have voted.

In the next chapter I present a liberal theory of liberty in which I argue that in order to understand liberty, we must understand power. The expectancy effects here discussed can be seen as the exertion of episodic power (Sattarov, 2019), and the result will potentially be a liberty-reducing interference that is manipulative and violates the requirement of respect for an individual's independence.

---

[1]This is related to the discussion of fairness in algorithms, and I refer to Binns (2018) for a discussion of fairness in machine learning based on political philosophy.

Privacy is a public good

The three threats may play in concert, but throughout this book it has also become clear that one of them is clearly the leader of the band. Surveillance is key to the increased effectiveness of nudging and algorithmic curation of data, and this clearly points to privacy as the key for facing any and all of these threats. In addition to showing that privacy is key for reducing each threat, it has also become clear that privacy is a special kind of good that cannot easily be solved by an individual approach to privacy control.

In daily parlance, a *public good* might refer to things that are good things for society and that most people appreciate—what Regan (1995) calls public and common values. When the term *common good* is used, it is often to illustrate the difference between private interests and the public good; the nuisance that individuals experience is, for example, outweighed by the public good. In the rest of this chapter I refer to privacy as a public good, but with a particular meaning different from the everyday understanding of the term. I turn to economic theory and privacy as a public good, or what Regan (1995) calls collective value.

Public goods are defined as *nonexcludable* and *nonrival*. Once the good is provided, no one can be prevented from enjoying it, and neither will one actor's use of the good impact other potential users' ability to enjoy the good (Barrett, 2007). Commonly used examples of such goods are lighthouses, national defense, and clean air. A key aspect of public goods is that they are prone to what is called *market failure*. When individuals do not bear the full cost of their actions, individually rational actions lead to collectively suboptimal outcomes. I may, for example, have a preference both for clean air and for driving a highly polluting car. I know that driving my polluting car will not deteriorate the air quality to such a degree that the air becomes unhealthy, so I believe that I can enjoy both clean air and my preferred car. It is thus individually rational for me to buy a gas-guzzler. However, if everyone reasons as I do and acts according to this reasoning, we will find ourselves in a collectively suboptimal situation without clean air.

This is known as a coordination problem, because individuals need some way to coordinate their actions so that they can jointly achieve a satisfactory balance between their desire to burn gasoline and for clean air. Similarly, individuals in a society must together find a satisfactory balance between privacy and other goods, such as ready access to fully personalized recommendations and entertainment. One particularly famous type of coordination problem is what game theorists refer to as the prisoners' dilemma.[2]

But how does clean air and prisoners relate to privacy? The tragedy of the commons is an example often used to explain the problematic nature of public goods (Hardin, 1968). When individuals have joint and unrestricted access to common resources, self-interested actors will be inclined to increase their exploitation of the resource, as they alone will get the added benefit, while the costs are divided by all actors. This, in turn, leads to the degradation of the resource. When privacy is seen as a public good plagued by externalities, we get *the tragedy of the privacy common*. Each individual will be inclined to accept a slight degradation of the common in order to derive an individual increase in other goods. For each individual, this is rational, and if only one individual did so, it would not necessarily degrade the common. The problem, however, is that all actors have the same incentive, and when many act on this incentive, the common is significantly degraded. Choi, Jeon, and Kim (2019) show how a market-based approach to privacy, with choice and full information, leads to suboptimal outcomes with excessive privacy loss. This, they argue, is due to information externalities and coordination failure—key characteristics of public good problems.

Relational leakages

Understanding *externalities* is crucial for understanding public goods. Externalities refer to a situation in which my actions have consequences for others. One example could be the pollution from a factory. Without any regulation of emissions, the factory owner's budget would not accurately reflect the cost others have to pay for their polluting activities, which would lead them to produce and pollute more than they would have done had they been required to take account of these external costs. An important privacy externality is that one individual's disclosed information can be used to infer information about other individuals (Choi et al., 2019). Barocas and Levy (2020) focus on three dependencies that help us understand the types of information leakage and privacy externalities that make privacy a public good: our *social relations* and our *similarities* to and *differences* from others. The first relates to the relational leakages discussed here, while the latter two are the topic of the next section.

---

[2]See Kuhn (2019) for a detailed analysis of this example.

Relational (or informational) leakages are also referred to as the network effect and related to the idea of networked privacy (Bannerman, 2019; MacCarthy, 2010; Ramachandran & Chaintreau, 2015). When some people are careless with their privacy, it has an impact not only on themselves but also on a wide range of people associated with them (Allen, 2011; Daughety & Reinganum, 2010; Regan, 1995). In addition, it has consequences for people both similar to and different from them. This regards one of the key ethical aspects of privacy management, as one person's self-disclosure has an impact on others (Morozov, 2013).

Recognizing our social natures, we see that information about me will by necessity involve some information about others as well. My relations and my social life are part of me, and even seemingly purely private information may be used to deduce information about, for example, my spouse or my friends. Knowledge of me involves knowledge of my relations (Fairfield & Engel, 2015). The result is that the threats stemming from Big Data cannot be assumed to be based on individual consent, and neither can they—even in theory—be effectively faced by individuals in isolation.

Even if one of my friends does not have a profile in social networks, surveillance agents might create "shadow profiles" for nonmembers, in order to (a) prepare their entry into the network and (b) better model the full social network, in order to improve their understanding of the actions of all involved (Horvát, Hanselmann, Hamprecht, & Zweig, 2012; Sarigol, Garcia, & Schweitzer, 2014). While surreptitiously gathering data on a person through indirect means is problematic, another challenge is related to how our similarities to others enables effective nudging and influence through algorithmic curation of data through the use of information about other people.

General profiles

Another aspect that makes privacy public is how the aggregation of various personal information lets surveillance agents build highly detailed generic profiles. When enough people like me willingly provide information about themselves, this information can subsequently be used to better understand and target *me*. This is related to the calls for group privacy, as knowledge of a group can be used to effectively target individuals in the group (Taylor, Floridi, & Van der Sloot, 2016). These issues are related to Big Data, and they show that the threats formed by Big Data need not be associated with detailed surveillance of each and every individual.

No one can be fully private in modern society, and some data exist on everyone. Some might have surrendered completely to the agents of data accumulation, and deep and comprehensive personality profiles will have been built. Others might only have left certain key variables, such as names, birth dates, location data, etc., in various datasets. They might not even have left them there themselves, as it could have been retrieved through various public sources of information. This leads to a situation in which there is always the option of superimposing highly detailed profiles onto individuals about which the surveillers only have limited information. We might all be unique—in principle—but we are also alike enough for this to be a problem. Fairfield and Engel (2015) use the example of deriving the chances of me getting cancer by using information about my brother. Even if we had no active social relations, similarities mean I am potentially hurt by his self-disclosure, for example, by having to pay a higher premium on my insurance. Furthermore, such similarities enable the nudging by expectations as discussed above.

It is, in short, impossible for me to be fully unknown in a world where everyone else is fully known. In public good terms, the public *bad* of zero privacy is also nonexcludable. If the public bad is provided, it is impossible for me avoid the harm that follows from it.

The difficulty of preserving privacy

One reason why economists and political scientists are so concerned with public goods is that they involve problems that are difficult to solve, and often require coercion or intervention of some other type. What, then, makes privacy so hard to preserve?

> *Imagine a group of rowers trying to propel a boat. Their speed depends not on the weakest rower, nor on the strongest, but on the efforts of all the rowers. Some global public goods likewise depend on the total efforts of all countries. Environmental issues are typically of this type. Pollution is determined by aggregate emissions, over-fishing by the fishing efforts of all countries.*
>
> **Barrett (2007, p. 74)**

One example of a public good is climate change, and this is a better parallel to privacy than one might first assume. There are four reasons, Barrett (2007) argues, why the optimal provision of certain public goods are hard to achieve. First, the dangers involved are somewhat diffuse and not immediate. Second, the consequences are different for different actors. Some will be harmed by a lack of privacy, while others will benefit. With climate change, the worst-off are the ones least equipped to combat the problem. The same might be the case with a loss of privacy. Third, protecting privacy will have consequences for our political and economic systems, and we must be prepared to forego certain opportunities for economic growth and innovation. Finally, and centrally, protecting privacy requires the *aggregate* effort of the actors involved. It is an essential feature of this particular public good that it does not require that *all* participate—only that *enough* do.

Diffuse effects and consequences, combined with limited understanding of the broader and long-term consequences of privacy loss (Ackerman, Darrell, & Weitzner, 2001; Nissenbaum, 2011), seem to be two most important factors that explain why we fail to provide the public good of optimal levels of privacy. The main factor, however, could be the uneven division of harm and benefit involved. Certain structures of harm and benefit give rise to rent seeking behavior, which is highly relevant to the issue of data protection. Whenever a market is not fully free, a competition for influence over whomever controls a market begins. Competition for favorable regulation, for example, is called seeking *rents* and is often (but not necessarily) associated with corruption, bribery, and outcomes that are not beneficial for society as a whole (Barrett, 2007; Krueger, 1974). The ones with the most to gain have an interest in combatting any regulation of government involvement that will change the current privacy regime. The people harmed, however, are many, and as each bears a relatively small burden of the cost, they will not have the same incentives to mobilize and act. This implies that tougher regulation and other forms of government intervention will be fought by those with much to lose. As Zuboff (2019) shows, some have a lot to lose, as there is great value in our information.

One consequence of the phenomena here discussed is increased academic and political attention to the role of big tech companies and issues of antitrust. While some see companies such as Facebook and Google as monopolies with a kind of structural power that leads to both economic and political harms, others might contend that these companies create new markets, novel and attractive services, and that making use of these services, and paying for them in part by the provision of data, is voluntary (Petit, 2020). While economic analyses and regulatory details of big tech is beyond the scope of this book, in Chapter 3 I discussed the superficial nature of the voluntariness involved in providing data, and in this chapter I have shown that since privacy is a public good, an individualized approach to privacy is fundamentally flawed and insufficient for facing the threats discussed.

Summary

While the three threats presented in Chapters 3, 4, and 5 are serious enough in isolation, this chapter has shown that the overall threat is exacerbated when they are seen in combination. In particular, it has been shown that nudging by sophisticated and data-driven expectations creates a novel threat due to the shaping power of algorithms based on this combination. Furthermore, the key roles played by surveillance and privacy—as an enabler of and potential safeguard against the threats—have been highlighted. Finally, I have argued that privacy is a public good which is highly susceptible to market failure if privacy concerns are left to individuals and a free market for privacy. This implies that unregulated markets for privacy will not be able to provide collectively optimal levels of the good. When individuals attempt to individually optimize their privacy levels, the result on a collective level will be that too little of the public good is provided. This explains the need for coordination and most likely government intervention.

References

Ackerman, M., Darrell, T., & Weitzner, D. J. (2001). Privacy in context. *Human-Computer Interaction, 16*(2−4), 167−176.

Allen, A. (2011). *Unpopular privacy: What must we hide?* Oxford: Oxford University Press.

Bannerman, S. (2019). Relational privacy and the networked governance of the self. *Information, Communication & Society, 22*(14), 2187−2202.

Barocas, S., & Levy, K. (2020). Privacy dependencies. *Washington Law Review, 95*, 555.

Barrett, S. (2007). *Why cooperate? The incentive to supply global public goods.* Oxford University Press.

Baruh, L., & Popescu, M. (2017). Big data analytics and the limits of privacy self-management. *New Media & Society, 19*(4), 579−596.

Bem, D. J. (1972). Self-perception theory. *Advances in Experimental Social Psychology, 6*(1), 1−62.

Binns, R. (2018). *Fairness in machine learning: Lessons from political philosophy.* Paper presented at the conference on fairness, accountability and transparency.

Bucher, T. (2012). Want to be on the top? Algorithmic power and the threat of invisibility on Facebook. *New Media & Society, 14*(7), 1164−1180.

Choi, J. P., Jeon, D.-S., & Kim, B.-C. (2019). Privacy and personal data collection with information externalities. *Journal of Public Economics, 173*, 113−124.

Daughety, A. F., & Reinganum, J. F. (2010). Public goods, social pressure, and the choice between privacy and publicity. *American Economic Journal: Microeconomics, 2*(2), 191−221.

Fairfield, J. A., & Engel, C. (2015). Privacy as a public good. *Duke Law Journal, 65*, 385.

Fox, J., Bailenson, J. N., & Tricase, L. (2013). The embodiment of sexualized virtual selves: The Proteus effect and experiences of self-objectification via avatars. *Computers in Human Behavior, 29*(3), 930−938.

Gillespie, T. (2014). The relevance of algorithms. In T. Gillespie, P. Boczkowski, & K. Foot (Eds.), *Media technologies: Essays on communication, materiality, and society.* Cambridge: MIT Press.

Hardin, G. (1968). The tragedy of the commons. *Science, 162,* 1243–1248.

Horvát, E.-Á., Hanselmann, M., Hamprecht, F. A., & Zweig, K. A. (2012). One plus one makes three (for social networks). *PLoS One, 7*(4), e34740.

Krueger, A. O. (1974). The political economy of the rent-seeking society. *The American Economic Review, 64*(3), 291–303.

Kuhn, S. (2019). Prisoner's dilemma. In E. N. Zalta (Ed.), *The stanford encyclopedia of philosophy* (2020 ed.). Spring.

Lanier, J. (2014). *Who owns the future?* Simon and Schuster.

MacCarthy, M. (2010). New directions in privacy: Disclosure, unfairness and externalities. *ISJLP, 6,* 425.

Merton, R. K. (1948). The self-fulfilling prophecy. *The Antioch Review, 8*(2), 193–210.

Morozov, E. (2013). The real privacy problem. *Technology Review, 116*(6), 32–43.

Nissenbaum, H. (2011). A contextual approach to privacy online. *Dædalus, 140*(4), 32–48.

Pearl, J. (2021). Radical empiricism and machine learning research. *Journal of Causal Inference, 8.*

Petit, N. (2020). *Big tech and the digital economy: The moligopoly scenario.* Oxford: Oxford University Press.

Ramachandran, A., & Chaintreau, A. (2015). The network effect of privacy choices. *ACM SIGMETRICS - Performance Evaluation Review, 43*(3), 59–62.

Regan, P. M. (1995). *Legislating privacy: Technology, social values, and public policy.* Chapel Hill: University of North Carolina Press.

Rosenthal, R. (1994). Interpersonal expectancy effects: A 30-year perspective. *Current Directions in Psychological Science, 3*(6), 176–179.

Rosenthal, R., & Rubin, D. B. (1978). Interpersonal expectancy effects: The first 345 studies. *Behavioral and Brain Sciences, 1*(3), 377–386.

Sarigol, E., Garcia, D., & Schweitzer, F. (2014). *Online privacy as a collective phenomenon.* Paper presented at the proceedings of the second ACM conference on online social networks.

Sætra, H. S. (2018). Science as a vocation in the era of big data: The philosophy of science behind big data and humanity's continued part in science. *Integrative Psychological and Behavioral Science, 52*(4), 508–522.

Sattarov, F. (2019). *Power and technology: A philosophical and ethical analysis.* Rowman & Littlefield.

Snyder, M., & Swann, W. B., Jr. (1978). Behavioral confirmation in social interaction: From social perception to social reality. *Journal of Experimental Social Psychology, 14*(2), 148–162.

Taylor, L., Floridi, L., & Van der Sloot, B. (2016). *Group privacy: New challenges of data technologies* (Vol. 126). Springer.

Yee, N., & Bailenson, J. (2007). The Proteus effect: The effect of transformed self-representation on behavior. *Human Communication Research, 33*(3), 271–290.

Zuboff, S. (2019). *The age of surveillance capitalism: The fight for a human future at the new frontier of power: Barack Obama's books of 2019.* New York: PublicAffairs.

7

Liberty in the era of Big Data

Introduction

What do we do when intuition tells us that our liberty is threatened, yet our theories fail to explain the threat we perceive? This is partly the case when attempting to get to grips with how Big Data threatens liberty. Three threats, individually and in concert, have been examined, but we must return to the concept that is purportedly threatened—liberty—in order to test whether classical theories of liberty are indeed incapable of explaining these threats. The time has come to gather the threads from the preceding chapters in order to explore the potential for formulating a liberal conception of liberty that satisfies three criteria. Firstly, I seek internal consistency, as the various conceptions of liberty examined in this book—negative and positive liberty, for example—cannot always be reconciled due their contradictory nature. Secondly, the concept must allow us to factor in the various forms of power exerted through Big Data. Thirdly, I seek a concept that is reconcilable with key traditional understandings of the concept as it has been known in the liberal theoretical tradition.

The latter requirement might appear to be somewhat strange. It is introduced as a requirement as much to test whether or not liberal theory is equipped to deal with the threats described in this book or if we must, as Cohen (2012) and Yeung (2017) suggest, move beyond liberal theory. I claim that one of the key features of liberal theory, and its most important contribution to human intellectual history, is its emphasis of respect for individuals and their individual liberty (Raz, 1986). If this is to be abandoned because this approach fails to account for new technologies and societal phenomena, it better be for good reason and based on a proper understanding of the potential of the individualistic approach to liberty. As such, seeking a description of liberty in the era of Big Data is both an examination of the purported shortcomings of liberal theory and of the potential for traditional theories of liberty to explain and make sense of issues

Big Data's Threat to Liberty. https://doi.org/10.1016/B978-0-12-823806-6.00002-8
Copyright © 2021 Elsevier Inc. All rights reserved.

such as nonphysical forms of power and interference and the various challenges created by surveillance and lack of privacy.

Some preliminaries must be settled before liberty itself is discussed, and these involve implementing a fuller understanding of power in order to more accurately describe interference, determining how to attribute actions and potential interferences to other individuals, and also the approach chosen to the question of whether or not liberties, choices, and ways of lives can be ranked, or must all be considered equal. After these preliminaries, I examine the potential for the emerging and restated conception of liberty as noninterference to explain and face modern technology-related threats to liberty.

Preliminaries

First of all, four important questions must be answered in order to arrive at an answer to the question of what liberty today entails:
- What is power?
- When does the exertion of power become illegitimate and constitute interference?
- How do we determine the source of interference and attribute exertions of power?
- Are some forms of interference, or some liberties, more valuable than others?

Power and attribution of action

Theorists of power often neglect to explain how power relates to liberty. The result is at times that everything is described as some form of power, while there is little to guide us as we want to determine which types of power exertion are problematic and which are not (Sætra, 2021b). Just as problematic is the case in which liberty is described at great length without an accompanying description of what sort of power constitutes interference. That is part of the problem described throughout this book, as some perceive liberal theory to be unable to explain the problems of Big Data because they, for example, rely on simplistic and restrictive notions of interference and liberal theory.

Combining a basic understanding of theories of power with the notion of noninterference allows us to solve this problem, without necessarily abandoning liberal theory, despite efforts by, for example, Yeung (2017) and Cohen (2012) to convince us to move to the nonliberal. While some criticism of liberal theory is clearly justified, I claim that much is also based on either misunderstandings or a lack of goodwill on behalf of the critics. Political theory, including liberal theory, is full of examples demonstrating awareness of how interference and force are not necessarily restricted to the physical varieties. I return to this in a later section, and here merely point toward some of the existing literature on power in its various forms.

Sattarov (2019), while not particularly concerned with coupling power with liberty, provides an up-to-date pluralistic conception of power that he also connects to technology. He draws on a wide variety of sources, including Amy Allen (1999), Mark Haugaard (2010), Stewart Clegg (1989), and Sheldon Wolin (2004), in his presentation of four types of power: episodic, dispositional, structural, and constitutive. By combining these in a pluralistic concept, just about every human action, can be said to be a form of power exertion.

This chapter seeks to develop a conception of liberty that allows us to determine what subset of power is liberty reducing and illegitimate. Simply saying that power is exerted is not particularly interesting if all actions entail an exertion of power. A parent exerts power over their children, the government exerts power over citizens, and a clever person exerts some form of power as they manage to get others to do as they please in life's various situations. The question of interest is: When is the use of power legitimate and when is it not? From the liberal position here developed, power is legitimate as long as it does not interfere with the liberties of others.

What the four types of power provide is a demonstration of the breadth of phenomena—physical and psychological—that can potentially be categorized as power, and thus interference. Episodic power is a relational form of power that characterizes the power one person has over another; the fleeting kind of power that exists in a particular context and will have to be reevaluated as the context and individuals involved change. Several types of episodic power are possible, including coercion, seduction, manipulation, persuasion, force, and authority (Sattarov, 2019). Coercion, manipulation, and persuasion have been examined in detail in Chapter 4, as these were described as psychological forces and forms of interference. Episodic power is crucial for understanding the threats posed by Big Data, and liberal theory is clearly open to taking such interference into account. Distinguishing between various forms of nonphysical influence is also part of the ongoing discussion of nudges powered by new technology, as seen both in this book and in Mills (2020).

A different form of power is nonrelational *dispositional* power (Sattarov, 2019). This involves focusing on the capacities and capabilities of the power-holder, without emphasizing relational or contextual inhibitors. I may, for example, have the power to lift my arm, lift an object, to wield a weapon, to build a sand castle, etc. As I emphasize interference with the ability to exercise one's liberty, episodic power is obviously relevant, but so is the analysis of interference with the use of dispositional power. When I say that I *can* access more information today than in any other period in history, this might refer both to a situation in which I actually do this and to a consideration of my hypothetical power to do so. Since the information certainly exists, and computers with internet access make the information available, the latter is certainly true. The first, however, might not be.

Various forms of influence, as described in the preceding chapters, might lead to a situation in which episodic power is exerted in a manner that renders my dispositional power irrelevant. Episodic power explains the relational aspect of liberty and interference, and it also helps to disentangle ability and capacity, which is the focus of dispositional power, form interference, and liberty (Crowder, 2013). This clearly links the discussion of power and liberty, as it is related to the choice of whether to focus on positive or negative liberty—or enabling or preventing conditions (Berlin, 2002a). Negative liberty entails a focus on what prevents us from making use of static dispositional powers, while positive liberty allows for a discussion of the dynamic nature of dispositional power.

Power has thus far been construed as a phenomenon related to individuals and individual (often dyadic) relations. Other concepts of power relate to the structural level, and this is referred to as systemic power (Sattarov, 2019). The structural power of Big Tech (Petit, 2020), for example, clearly has implications for liberty. However, the main focus of this book is how individual liberty is affected by Big Data, and I have chosen to emphasize the more direct relational exercise of power and interference, rather than focusing on the distribution of power and liberty between various groups. Analyzing systemic power allows us to determine the distribution of ableness and powerlessness, and this is highly relevant, but not a part of the discussion of how Big Data might theoretically be used by someone to exert power in a way inimical to anyone else's liberty. However, systemic power might involve the use of ideological power (Sattarov, 2019), and this again relates to the power of persuasion and shaping of minds discussed in this book. It is also fundamentally related to the questions of structural restraints on individuals and groups—women and men both—and how free choice is relatively meaningless unless we account for the upbringing and context in which certain choices make sense and are associated with a wide range of formal and informal consequences (Hirschmann, 2013). As interference is not restricted to physical power, this also relates to the notion of "soft power," discussed by Nye (2011) in the context of international relations.

Finally, there is what Sattarov (2019), building on Foucault in particular, calls constitutive power. This entails the power to construct, shape, and build the very foundation of human's selves. This is related to the shaping power of algorithms, as discussed in the previous chapter. Of particular interest is how incorporating such phenomena necessarily involves moving away from a fully static conception of individuals—their dispositional power included. As power can be used to change a person, their dispositional power is affected as a consequence, as will their preferences and desires regarding the use of such powers be. This is reminiscent of Cohen's (2012) discussion of the role of privacy for protection and the growth and integrity of the "postliberal" self. However, her critique of liberal theory and call for a move to nonliberal theory is premised on an assumption that liberal theory is incompatible with a dynamic conception of the self. This is thoroughly contested in this book, as the liberal argument presented clearly does allow for the analysis of an accommodation of such change in individuals.

In the following sections, all four types of power are acknowledged, and I follow Sattarov (2019) in his proposal that we use a pragmatic compound concept of power as a way to analyze various perspectives and aspects of power. However, my main focus is not to determine what is power and not, but which types of power exertions are legitimate and which are illegitimate and thus constitute liberty-reducing interference.

Attributing interference

I consider the actions of humans as the only relevant source of interference. As previously discussed, I have dismissed Sunstein's notion that nature can nudge, and I also reject the notion that machines or algorithms can be responsible for actions. This involves a rejection of the idea that highly advanced machines create some form of responsibility-gap when they approach certain levels of autonomy and complexity (Gunkel, 2017; Matthias, 2004). While a *veil of complexity* obscures the human source of machine actions (Sætra, 2021a), I posit that the responsibility for such actions nevertheless remains squarely placed in human beings. This is based on the notion that machine action is ultimately the direct result of various human choices of programming and design, and regardless of the unpredictability thus introduced the choice to develop and deploy such unpredictable machines comes with responsibility for the actions that follow (Sætra, 2021a).

But how does this relate to liberty and power? Four approaches to the attribution of actions were introduced in Chapter 2: causal responsibility, moral culpability, moral responsibility, or intentionality. A causal responsibility approach to obstacles to freedom might seem attractive, as this attributes an effect to people who have in some way taken part in the causal chains that led to this effect. However, the problems related to tracing causal contributions, for example, to the workings of the nudging that occur on a social platform are so great that such a principle becomes completely unworkable in practice. Furthermore, considering the various ways in which innocuous actions affect the actions of others, it seems unsatisfactory to attribute responsibility for all that play a causal role, regardless of their intentions or their knowledge of the consequences of their actions.

I rely on the approach that entails considering interference liberty reducing only when some person can be held morally responsible for it—when the action is morally attributable (Carter, 1999). Such an approach requires that the interference will *either* be intentional or that the person is negligent. Negligence here involves that they could and should have considered the consequences of their actions, and even if they didn't, they will be assumed to have acted on such knowledge of the consequences. Intentions matter, but they are not all that matters. The same goes for causality. Being involved in causing the possibility for or occurrence of some consequence is a necessary, but not sufficient, condition of responsibility.

As to the power exerted through Big Data, the approach here described has clear consequences for further research on who should be held responsible for the liberty lost, but my main goal here is to determine how and whether liberty is lost. The approach suggests that whenever people design systems that have the potential to influence actions, and employ them in ways in which they must be expected to actually influence actions, the potential for human-attributable interference exists. However, it still remains to determine what sort of influence and power exertion is deemed interference.

Perfectionism and the ranking of values

A key controversy in the philosophy of liberty is whether or not we can evaluate and rank the various options and choices available to individuals. Some like Carter (1999) have denied the possibility of valuing *specific* liberties, and as we saw in Chapter 2, he focuses on liberty as a counting game in which overall liberty is simply the ratio of actions available to me and all potentially available actions (available or not). Such an approach would involve a completely different analysis of the effects of Big Data than the one described in this book, as any specific potential loss of liberty would mainly be problematic if it was not simultaneously accompanied by new available actions or choices. The new possibilities provided by new technologies could thus easily be argued to increase overall freedom.

However, a completely different approach involves starting the analysis of liberty and interference with the idea that certain options and actions are more important or fundamental than others. This is the called a *perfectionist* approach to liberty (Raz, 1986), and while freedom of choice and the number of choices matter, *which* choices are available matters even more. For example, the liberty of killing a person I dislike and the liberty of expression do not have to be considered equal in a moralized or perfectionist theory of freedom. There is a foundational incommensurability in terms of values, but the perfectionist theory here presented entails that some values can be said to be of importance and others less so, and that certain forms of interference are legitimate and not liberty reducing (List & Valentini, 2016). Carter (1999) and List and Valentini (2016) argue in favor of nonperfectionism, while Pettit (1997) and Raz (1986) exemplify the perfectionist approach.

The current purpose is not to determine which approach is most correct. In fact, this might be an irrelevant question, and I propose that the two approaches are not incompatible, but that they are different types of tools for analyzing different types of liberty. This is analogous to the pragmatic approach to power, in which different theories allow for different types of insight into what power is. Nonperfectionist theory will, for example, consider *any* kind of interference liberty reducing (List & Valentini, 2016), and while this is superficially true, a political theory of liberty might benefit from distinguishing between legitimate and illegitimate forms of interference. For example, Bastiat (1998) and other liberals in favor of limited government recognize that the law, and its enforcement, *protects* liberty, despite its coercive nature.

List and Valentini (2016) refer to perfectionist liberty as moralized liberty, and they argue that Berlin was a proponent of nonmoralized liberty, but that his conception of liberty suffers from not including Pettit's and the republican tradition's considerations of the *potential* for interference. They thus seek to create what they perceive as the best of both worlds, in a nonmoralized form of robust or guaranteed liberty. As they explicitly seek to examine the freedom of actions, and not the value of those actions, their theory is of little use if the goal is to examine which forms of power exertions are legitimate and if one holds the view that certain liberties are far more valuable than others. In addition, the combination of nonperfectionism with the republican notion of nondomination where possible-but-improbable constraints are considered creates an impossible situation in which there is no liberty (Carter & Shnayderman, 2019).

While I propose a moralized account of liberty, it is based on the notion that certain liberties are foundational, and that it makes little sense to weigh foundational liberties against fripperies as if they were equal. Berlin, and the negative liberty he championed, is often caricatured and turned into a strawman, not only by theorists from nonliberal theory but also by List and Valentini (2016) and others who use him as an example of a theorist of "pure" non-moralized noninterference. While such readings of Berlin can be defended on the basis of his "Two Concepts of Liberty" (Berlin, 2002b, p. 38), in a later introduction to this article he attempts to correct some of the "astonishing opinions which some of [his] critics have imputed" to him. He explains that the misconceptions stem from him not having stated what he had thought to be obvious to all, for example, that unrestricted laissez-faire might result in a lot of theoretical liberty, but also great dangers to negative liberty—which entail such fundamental liberties as basic human rights, free expression, and free association, which he deems to be prerequisites of democracy (Berlin, 2002b). Negative liberty requires systems in which it can be exercised, and without such systems, and the basic and foundational liberties that characterize such systems, the value to individuals of theoretical liberties without any means to avail themselves of them is nil. Berlin also supports the notion that government intervention to protect negative liberty is obviously legitimate, and he had assumed this to be self-evident, and also states clearly that this is understood and clearly stated by most prominent liberal theorists, including Tocqueville, Mill, and Benjamin Constant (Berlin, 2002b).

Berlin, unlike many of his readers, considered negative and positive liberty to be complementary— not mutually exclusive—conceptions of liberty (Crowder, 2013). While he emphasized the value of negative liberty, and the dangers of certain varieties of positive liberty, nowhere does he argue that negative liberty in some minimalist form is all that matters. That is also the liberal position here restated—one that clearly emphasizes noninterference, but simultaneously allows for a consideration of foundational liberties and the very preconditions for liberty. There is no shortage of philosophers that have critiqued Berlin's distinction between positive and negative liberty (Dimova-Cookson, 2013), arguing that their position lies between positive and negative liberty (Pettit, 1997), or in favor of the need to account for both kinds (Raz, 1986). I here argue that Berlin does the same, and only the strawman version of Berlin can be said to focus exclusively on negative liberty.

Without being able to go fully into the details of rights and the source of liberties, the following is based on the liberal notion that liberties are valued according to their importance for allowing a wide range of individuals to live what they themselves consider good lives, as partly discussed in relation to paternalism in Chapter 4. The preservation of negative liberty is thus the goal of liberty reductions, and paternalist approaches aimed at enforcing particular conceptions of the good life are considered illegitimate.

Some liberties relate to the potential for pursuing ultimate or foundational goods (Raz, 1986), while other liberties may indeed be considered fripperies. Some basic liberties can be seen as the source of all liberty, and thus be foundational, while others—such as the availability of a wide range of ice cream flavors—are at best derivative liberties. While the foundations of liberty are important, it is also important not to conflate liberty and other political values, such as fairness, equality, or justice (Berlin, 2002b).

This approach allows us to evaluate specific freedoms, and we might even say that some liberties are of great importance, while others are worthless, or even detrimental, to liberty in general. Raz (1986, p. 409) argued that negative liberty is insufficient, as it is relatively blind to an individual's "pervasive goals, projects and relationships" and thus cannot help us to decide what liberties are actually important. The basic component of this approach is Berlin's (2002b) value pluralism, and it entails a rejection of political rationalism (Orlie, 2013). Various components of positive liberty can be legitimate individual values and goals, but these cannot be made *political goals* through paternalistic policies aimed at helping—or forcing—individuals to understand or actually realize them (Myers, 2013; Orlie, 2013).

Interference: beyond the physical

The idea that liberal theory is only focused on physical power and interference has been rejected, and such a rejection is neither speculative nor controversial, but based on a very basic reading of classic theories of liberalism, liberty, and power. Liberal theory can account for moral or social power (Tully, 2013), and also the kind of power exerted through *structures*, and the diffuse power exerted through technology (Coole, 2013). Similarly, the notion that freedom of choice is all that matters for a liberal is also dismissed, as liberal theory can clearly be based on a perfectionist conception of liberty.

As the nonphysical aspects of power are accepted, this relates the discussion of liberty directly to an emphasis on the value of *autonomy*, and many attempts to marry the pursuit of negative liberty with an account of agency have been proposed (Christman, 2013). In fact, it is not possible to determine what is interference without a form of account of an individual's agency, and Berlin also recognized this as he refers to individual's desires, abilities, preferences, that "codefine the area of noninterference" (Tully, 2013, p. 31). Autonomy might require that a person be the "part author of his own life" and that their life is "marked not only by what it is but also by what it might have been and by the way it became what it is" (Raz, 1986, pp. 204, 369). For a person to be autonomous, we can require that certain conditions are fulfilled.

One proposed set of conditions of autonomy is that a person has a certain level of mental abilities, an "adequate" range of options, and independence (Raz, 1986). The first was partly discussed in relation to psychological force and weakness in Chapter 4, and I mainly focus on the latter two conditions of autonomy here, and in particular the third. Independence is particularly central in determining when an exertion of force is interference of a kind that is at odds with the independence of others. One specific requirement for a person to be independent is that they are not subject to the coercion or manipulation of others (Christman, 2013; Raz, 1986). These two concepts relate closely to the preceding discussion of power, and will be explored in more detail below.

Regarding an adequate range of options, this will be taken as a demand for real choices beyond those available to a person desperate for survival, for example. However, it also requires that a person living a life of luxury has meaningful options available to them, and this is related to the evaluation of which choices or liberties are foundational and important. It must here be noted that options associated with serious social sanctions are not considered fully available, as the moral power of others can constitute a form of interference. This is of particular relevance to the question of the tyranny of perceived opinion, discussed in Chapter 5. There are many forms of life which are incompatible with autonomy, as autonomy requires societies in which meaningful options are not only theoretically available but also that enough of these are not associated with the significant negative consequences that render the room for individual choice insufficient (Raz, 1986). Privacy is here regarded as a key safeguard of this required space, but the condition of adequate choices also places certain demands on the social context in which full privacy will never be a real or desirable alternative.

The third condition for freedom is independence, which requires freedom from coercion and manipulation (Raz, 1986). Coercive and manipulative forms of interference are key questions dealt with in Chapters 4—6. In relation to nudging, I discuss both how the use of psychological force may be coercive, and the fact that nudging involves techniques which may be considered manipulative. In Chapters 5 and 6 I focus more on how agents— deliberately or accidentally through algorithmic curation of information—manipulate the information we receive, and consequently manipulate our world views and our conditions for developing the basic faculties and abilities mentioned by Berlin (2002b) and Cohen (2012).

What does the condition of independence really entail? Coercion and manipulation must be defined, but I begin with discussing the possible connection of these concepts with the idea of threats—one of the nonphysical ways of exerting power.

Coercion, threats, and manipulation

Coercion is often associated with the use of physical force, but it is here used more broadly to describe any use of force that effectively removes alternatives of action. Carter (1999) has stated that threats only reduce liberty when they are credible and then enacted. However, people do not act on the basis of certain knowledge of future consequences, but on the consequences they believe will come to pass should they perform some action (Raz, 1986). A bluff that is believed will thus affect behavior just as much as communication of true intent. When I threaten someone, I aim to give them a reason to act in a certain way. When the threat is credible it *is* such a reason (Raz, 1986). Since people act on reasons, their actions are clearly affected by the external imposition of reasons, even if such threats do not make actions physically impossible.

Not all liberals would agree that an action performed under threat is necessarily unfree. Hobbes, for example, provides examples of how actions taken to avoid worse consequences—even when these worse consequences are imposed intentionally by others—are not unfree (Hobbes, 1946). If I make a choice to avoid the scorn of others, for example, this can be seen as a deliberate and free choice. However, another perspective is more in line with the liberal theory that emerges in this book. It is possible to argue that *duress* relieves an actor of obligation and that the responsibility for actions taken under duress might be transferred to others if the duress is morally attributable to them (Raz, 1986). The Hobbesian gunman—in a typical *"give me your money or I'll kill you"* scenario—threatening to shoot a person if they do not hand over their wallet creates a *forced choice* (Raz, 1986). Berlin also speaks of freedom as consisting in not making forced choices, and seems to relate this to the types of scenarios just discussed (Hirschmann, 2013).

The fundamental issue is that a coerced, or forced, person is acting against their will. We must often act against our will, so this in itself is not necessarily a problem. It becomes a problem, however, when the reason we act in spite of our will is some form of interference morally attributable to someone else. Coercion involves subjecting the will of another to one's own (Raz, 1986), and this is a clear violation of autonomy. We might also note that people may have comprehensive goals, and when we force a person into a situation where all but *one* alternative involves abandoning his goal, this is also understood as coercive (Raz, 1986). This is of particular interest with regard to the discussion in Chapter 3, where I argue that a lack of alternatives to living a life where one must interact with the phenomenon of Big Data is a threat to freedom.

In this context we must distinguish between a general coercive pressure, which can be morally attributable, and what is called threats, which is here exclusively used to describe intentional actions taken to change another's actions. Both can be liberty reducing. I might be threatening without this being my intention, but this constitutes a different sort of interference than the threats I here discuss. If I use algorithms to change a person's view of the world, without it being my intention to change his actions in a certain way, I am not reducing liberty by threat, even if I might be morally responsible for reducing their liberty through undermining their autonomy by manipulation, which will be discussed shortly.

If we abide by the principle that one is only responsible for one's own actions, we *do* become responsible for the actions of others if "one is responsible for inducing another to act" by deception of "or an intention to cause harm of to exploit" (Raz, 1986, p. 95). To be clear, we are not responsible if we persuade another person to act on reasons different from those he used to hold, if our method is that of rational persuasion (Raz, 1986). The difference between nudging and rational persuasion is discussed in more detail in Chapter 4.

Coercion by threat is thus conceived of as possible, which implies that we are not limited to discussing physical coercion when evaluating possible interference by other people. Some even state that coerced persons are both unfree and not responsible for their actions, as "the coerced are being controlled by another in a way akin to physical coercion" (Raz, 1986, p. 149). This is somewhat similar to the development from legislating only physical interference to acknowledging that we must protect *more*, as described by Warren and Brandeis (1890) in their seminal work *The Right to Privacy*. List and Valentini (2016) similarly argue that laws, and not simply physical actions, stand in need of justification, and are thus potential sources of interference.

These conclusions should not come as a surprise to those acquainted with liberal political theory, and the fact that it must be stated suggests that liberal theory is often both caricatured and misunderstood. Three suggestive examples discussed in this book are Goodwin (2012), Cohen (2012), and Yeung (2017). Also theorists of liberty have argued that phenomena such as psychological manipulation and coercion of the will do not infringe upon negative liberty (Carter, 1992), but others, in line with the arguments of this book, are of the opinion that this is needlessly restrictive, and that Berlin himself also at times seems to acknowledge this (Coole, 2013). When it is realized that liberal theory is compatible with an understanding of individuals as situated beings (Christman, 2013), as negative liberty being more than physical intervention (Coole, 2013), and that liberal theory is not in any way identical to what is disparagingly referred to as "neoliberalism" (Orlie, 2013), it becomes apparent that liberal theory might be able to account for some, if not all, of the threats discussed in this book after all.

While coercion is usually assumed to be liberty reducing, manipulation is treated in a variety of ways by liberal theorists. While coercion involves interference with a person's options, manipulation "perverts the way that person reaches decisions, forms preferences or adopts goals" (Raz, 1986, pp. 377−378). Some speak of a form of "micro-power" that is not deliberate, intentional, or visible, and power that "manipulates the will" (Coole, 2013, p. 200). Like coercion, manipulation is regarded as a clear invasion of autonomy, and is thus liberty reducing, as the absence of both manipulation and coercion is a condition of autonomy. Manipulation consists in *perverting* the way a person makes decisions in order to subject the will of the agent to that of the manipulator. This can be done through manipulating the decision-making process, preference formation, or goal adaption (Raz, 1986). The perversion of our decision-making processes is emphasized in Chapter 4, and further elaborated in Chapter 6. It involves, for example, appeals to known personal biases or irrational proclivities, instead of appealing to a person's rational decision-making faculties. While we constantly make decisions under the influence of such irrational mechanisms, I focus on the morally attributable exploitation of such mechanisms. The manipulation of, or interference with, the formation of preferences and goals, through algorithmic curation of information, is emphasized in Chapters 5 and 6. The application of the techniques of nudging is here understood to be an intentional action, while the broader exertion of algorithmic power can be problematic despite there being no specific intentions related to the threats I here described. Even if the intentions of the actor who employs algorithms that change our view of the world were neutral, or good, they are morally responsible for the consequences that follow if our liberty is reduced as a result of their actions.

Interference and privacy

Privacy is related to liberty in various ways, and if Warren and Brandeis (1890) are correct in that privacy incursions cause pain and distress "far greater than could be inflicted by mere bodily injury," one specific connection with liberty is uncovered. Great harm is clearly something we can potentially prevent others from inflicting in the name of liberty.

I could also argue that a lack of privacy reduces liberty because I lose the right to do a certain action X at home while not being observed. However, if we are playing Carter's (1999) counting game, I could add that the lack of privacy simultaneously introduces the possibility of doing the same action, at home, while being observed. For example, if I am under complete surveillance, I lose the option of reading a good night story for my children in private, while I gain the option of reading while being observed against my will. This is a different option from voluntarily live streaming the reading session online—an option I would have had regardless of the availability of privacy.

As discussed, for the lack of privacy to entail a loss of liberty, we must be willing to rank the various options that exist and argue that privacy is a foundational good that cannot be traded for other, even if they are more numerous, "lesser" goods. As soon as we incorporate requirements of autonomy and independence, another link between privacy and liberty becomes apparent: privacy is required for providing Berlin's "minimum area" of liberty in which an individual may form and develop the faculties and skills required for a human existence. For a liberal theory concerned with the preservation of negative liberty, privacy is thus clearly of great value, and there is no need to move toward a comprehensive theory of positive liberty in order to justify this valuation.

The notion of privacy grew forth alongside the notion of the individual. Before the idea of individuals as meaningfully isolated bearers of rights and duties, the division between the private and public made little sense. Berlin (2002b, p. 176) states that privacy and the idea of "the area of personal relationships as something sacred in its own right" derive from the negative conception of freedom and individual liberty, and that the loss of it "would mark the death of a civilisation, of an entire moral outlook."

Privacy as a precondition

One reason for liberals to defend privacy is the risk of abuse, theft, or other forms of exploitation of personal data. However, there is more, and positive and negative liberty come together in defense of privacy as a precondition for liberty because it defends the private area necessary for developing and safeguarding individual development.

What is to be developed? Cohen (2012) mentions the capacity for self-determination and self-development, in addition to the development of critical perspectives. Berlin noted the need to develop the faculties required to "pursue, and even to conceive, the various ends which men hold good or right or sacred"—a human being's natural faculties (Berlin, 2002b, p. 171). These conditions are not merely the preconditions of liberty but also *constituent elements* of it (Tully, 2013). Privacy creates and protects the boundaries around the self that make it possible to develop the capacity for self-determination and self-development (Cohen, 2012), and these are clearly valuable capacities also for liberal theorists, as liberal theory is concerned with the provision of the conditions of liberty, and not merely the abstract analysis of theoretical forms of negative liberty. The way privacy is connected to freedom through the need for a sphere in which we are not observed or interfered with is the main topic of Chapter 5.

While some claim that the right to privacy draws the line between what is private and what is political, Raz suggests that the importance of such rights is also to protect "a collective good, an aspect of a public culture" (1986, p. 256). They are not considered to be of great importance because they determine what is beyond politics. Instead, they are of great importance, and thus a political issue, due to their importance in the provision of the public goods required to provide the conditions for liberty. In this context it is also important to remember that privacy is a public good.

The implications for liberal policy

With this understanding of liberty, we are finally able to understand how all the phenomena discussed threaten liberty, and in closing I will briefly examine the political implications of the liberal theory here restated.

Liberalism is here understood as the attempt to foster and develop societies that promote liberty and value pluralism. This implies that liberalism provides a wide range of reasons to take issue with Big Data, as it has been shown that it threatens fundamental liberties and the potential for realizing a liberal society. Society and the opportunities of individuals to live their lives as they desire are constantly changed by social, economic, and technological processes (Raz, 1986). Over 125 years ago, Warren and Brandeis (1890) noted that as technologies evolve, our governments and laws must also change. Alan Westin (1967), in *Privacy and Freedom*, similarly deals with how privacy is challenged by new technologies. Technology *can* be seen as a force we consider to be uncontrollable. However, I do not adhere to such a view. Næss (1989) describes the somewhat opaque nature of the development of techno-economic systems. If we do not dare to question and control technology, we might soon find ourselves in a situation in which "[t]he cog-wheels have drawn us into the very machine we thought was our slave" (Næss, 1989, p. 24). Technology is not, and will never be, neutral, and it must be subject to "evaluation in normative systems" (Næss, 1989, pp. 94–95). One such evaluation is the current examination of Big Data and liberty.

Liberty requires the satisfaction of certain conditions in order to prevail. If we agree that liberty as here conceived is morally valuable, and a right all people have, our governments have a clear mandate to provide and safeguard the conditions of liberty. Government has an essential role to play when individuals and private companies threaten the liberty of others, and we must call upon our governments to be the guarantors of liberty, instead of seeing their reluctance to interfere as liberty increasing. Liberal theory is clearly compatible with the notion that noninterference should be accompanied by a certain degree of non*domination* (Pettit, 1997), and the protection against potential arbitrary power is the very foundation of the liberal ideal of the rule of law and individual rights.

Protection against arbitrary power does not only consist in protection from the power of tyrants and government. Today, "private corporations … have as much, if not more, power than many public authorities" (Raz, 1986, p. 4). While this might have been true when written in 1986, I argue that it is an even more important insight today. While focusing on political institutions is important, it is also imperative that we recognize private power as one of the most important justifications for advocating in *favor* of political power.

The theory of liberty here presented provides us with a coherent account of why the government should limit and regulate private enterprise when it inhibits personal autonomy, and while so doing it must also refrain from using the possibilities that technology provides in a similar way itself. Like Mill, we can use the harm principle to explain why liberals may justifiably coerce to prevent harm (Raz, 1986). While such a view of liberty places government in a more active role than some other forms of liberalism, it also places strict boundaries on government action, and there is a recognition of the fact that power both corrupts and is corruptible (Acton, 1907; Raz, 1986). We should not *trust* government, but hedge and fence its power, while giving it enough authority to create the conditions for autonomy, and thus good lives (Raz, 1986).

As established in the previous chapter, privacy is central for the provision and protection of liberty. Firstly, privacy is central with regard to having the space to ensure the development of basic human capabilities and fostering independence. Moral pressure and pressure of conformity are inimical to liberty, and privacy is one safeguard against such phenomena. Secondly, privacy helps protect individuals against targeted efforts to manipulate and coerce through the perversion of our decision-making situations and processes. When considering independence, we do not simply evaluate the current state of a person, but also how this current state came about.

Moreover, we see how we can understand the nonphysical aspects of coercion and manipulation, and the fact that such efforts are clear violations of the conditions of liberty, and thus inimical to liberty. This applies both to efforts to nudge and prod us, and to the general control and curation of information by actors beyond political control.

Liberty also requires that we have adequate options to choose from. One role of government might then be to restrict the growth of what Zuboff (2019) labels *surveillance capitalism*, as this growth may take us into a situation where we no longer have the option of living outside the gaze of those that surveil us.

Summary

Liberty is a broad and contested concept, and in this chapter I have dealt with selected key preliminaries and related concepts. The result has been a restated liberal conception of liberty, in which the role of power, the attribution of actions, and the need to rank values are highlighted. In addition, I have argued that interference is more than physical obstruction, and that the conditions of liberty, privacy being a key component, are essential for any liberal theory that aims to understand the effects of Big Data.

The theory of liberty here presented shows how liberty is affected by Big Data, and also highlights how we might go about it should we wish to protect and preserve it. If we do perceive the threats described in this book as real, and if a simplistic and restrictive conception of negative liberty is unable to explain these threats without great contortion and strain, this is a good reason to move beyond negative liberty, rather than merely concluding that the threats are not real. Furthermore, I have implied that Berlin himself clearly recognized that both aspects of positive liberty and other political values are of great importance, and that he himself states that we must in fact also see beyond negative liberty.

Finally, liberal theory and liberalism are not incompatible with coercion, as the preservation of individual liberty requires laws and constraints of various types. By establishing that privacy is a public good it has also been shown that the harm principle can be invoked to coerce privacy in accordance with liberal theory.

References

Acton, B. J. E. E. D. (1907). *Historical essays & studies*. limited: Macmillan and Co.

Allen, A. (1999). *The power of feminist theory: Domination resistance solidarity*. Boulder: Vestview Press.

Bastiat, F. (1998). *The law*. Irvington-on-Hudson: Foundation for Economic Education.

Berlin, I. (2002a). *Liberty*. Oxford: Oxford University Press.

Berlin, I. (2002b). Two concepts of liberty. In H. Hardy (Ed.), *Liberty*. Oxford: Oxford University Press.

Carter, I. (1992). The measurement of pure negative freedom. *Political Studies, 40*(1), 38–50.

Carter, I. (1999). *A measure of freedom*. Oxford: Oxford University Press.

Carter, I., & Shnayderman, R. (2019). The impossibility of "freedom as independence". *Political Studies Review, 17*(2), 136–146. https://doi.org/10.1177/1478929918771452.

Christman, J. (2013). Freedom, autonomy, and social selves. In B. Baum, & R. Nichols (Eds.), *Isaiah Berlin and the politics of freedom: 'Two concepts of liberty' 50 years later* (Vol. 50, pp. 87–101). New York: Routledge.

Clegg, S. R. (1989). *Frameworks of power*. Sage.

Cohen, J. E. (2012). What privacy is for. *Harvard Law Review, 126*, 1904.

Coole, D. (2013). From rationalism to micro-power: Freedom and its enemies. In B. Baum, & R. Nichols (Eds.), *Isaiah Berlin and the politics of freedom: 'Two concepts of liberty' 50 years later* (Vol. 50, pp. 199–215). New York: Routledge.

Crowder, G. (2013). In defense of Berlin: A reply to James Tully. In B. Baum, & R. Nichols (Eds.), *Isaiah Berlin and the politics of freedom: 'Two concepts of liberty' 50 years later* (Vol. 50, pp. 52–72). New York: Routledge.

Dimova-Cookson, M. (2013). Defending Isiah Berlin's distinctions between positive and negative freedoms. In B. Baum, & R. Nichols (Eds.), *Isaiah Berlin and the politics of freedom: 'Two concepts of liberty' 50 years later* (Vol. 50, pp. 73–86). New York: Routledge.

Goodwin, T. (2012). Why we should reject 'nudge'. *Politics, 32*(2), 85–92.

Gunkel, D. J. (2017). Mind the gap: Responsible robotics and the problem of responsibility. *Ethics and Information Technology*, 1–14.

Haugaard, M. (2010). Power: A 'family resemblance'concept. *European Journal of Cultural Studies, 13*(4), 419–438.

Hirschmann, N. J. (2013). Berlin, feminism, and positive liberty. In B. Baum, & R. Nichols (Eds.), *Isaiah Berlin and the politics of freedom: 'Two concepts of liberty' 50 years later* (Vol. 50, pp. 185–198). New York: Routledge.

Hobbes, T. (1946). *Leviathan*. London: Basil Blackwell.

List, C., & Valentini, L. (2016). Freedom as independence. *Ethics, 126*(4), 1043–1074. https://doi.org/10.1086/686006.

Matthias, A. (2004). The responsibility gap: Ascribing responsibility for the actions of learning automata. *Ethics and Information Technology, 6*(3), 175–183.

Mills, S. (2020). Nudge/sludge symmetry: On the relationship between nudge and sludge and the resulting ontological, normative and transparency implications. *Behavioural Public Policy*. https://doi.org/10.1017/bpp.2020.61.

Myers, E. (2013). Berlin and democracy. In B. Baum, & R. Nichols (Eds.), *Isaiah Berlin and the politics of freedom: 'Two concepts of liberty' 50 years later* (Vol. 50, pp. 129–142). New York: Routledge.

Næss, A. (1989). *Ecology, community and lifestyle: Outline of an ecosophy*. Cambridge University Press.

Nye, J. S. (2011). *The future of power*. New York: Public Affairs.

Orlie, M. A. (2013). Making sense og negative liberty: Berlin's antidote to political rationalism. In B. Baum, & R. Nichols (Eds.), *Isaiah Berlin and the politics of freedom: 'Two concepts of liberty' 50 years later* (Vol. 50, pp. 143–154). New York: Routledge.

Petit, N. (2020). *Big Tech and the digital economy: The moligopoly scenario*. Oxford: Oxford University Press.

Pettit, P. (1997). *Republicanism: A theory of freedom and government*. Clarendon Press.

Raz, J. (1986). *The morality of freedom*. Oxford: Clarendon Press.

Sætra, H. S. (2021a). Confounding complexity of machine action: A Hobbesian account of machine responsibility. *International Journal of Technoethics, 12*(1). https://doi.org/10.4018/IJT.20210101.oa1.

Sætra, H. S. (2021b). *Review of power & technology*. Prometheus.

Sattarov, F. (2019). *Power and technology: A philosophical and ethical analysis*. Rowman & Littlefield.

Tully, J. (2013). "Two concepts of liberty" in context. In B. Baum, & R. Nichols (Eds.), *Isaiah Berlin and the politics of freedom: 'Two concepts of liberty' 50 years later* (Vol. 50, pp. 23–51). New York: Routledge.

Warren, S. D., & Brandeis, L. D. (1890). The right to privacy. *Harvard Law Review*, 193–220.

Westin, A. F. (1967). *Privacy and freedom*. New York: IG Publishing.

Wolin, S. S. (2004). *Politics and vision: Continuity and innovation in western political thought*. Princeton: Princeton University Press.

Yeung, K. (2017). 'Hypernudge': Big Data as a mode of regulation by design. *Information, Communication & Society, 20*(1), 118–136.

Zuboff, S. (2019). The age of surveillance capitalism: The fight for a human future at the new frontier of power: Barack Obama's books of 2019. New York: PublicAffairs

8

Conclusion

This book has detailed the relationship between Big Data and liberty, with a particular focus on unpacking the notion of liberty. Liberty, or freedom, is mentioned in a wide array of books and articles that analyze the consequences of technology, but most of the times the content of the concept is not made explicit. This is a problem for the concept of liberty in general, as it is usually praised and put forth as an ideal, while the fact that it is a deeply contested, and at times ill-understood, concept, is often glossed over or neglected (Baum & Nichols, 2013). While much space has been devoted to distinguishing between various forms of liberty, the main goal has not been to determine which conception is best, or most correct. Rather, it has been to use the different conceptions of liberty to explain how Big Data influences liberty, and also how liberal theory is able to account for such threats. This is in line with the suggestion that we should not seek to fix the meaning of liberty once and for all, but to use the insight derived from the debates about liberty to illuminate and explain contextualized and practical challenges (MacGilvray, 2013).

I have drawn three partial conclusions in the chapters that detail isolated threats (Chapters 3–5). The first is that surveillance is a threat to liberty, as privacy is a necessary condition for liberty. I argue that even passive observation constitutes a threat, and that this is true even if one adopts the somewhat limited negative conception of liberty. The second conclusion is that Big Data and technologies of personalization have the potential to turn the famous nudge into a shove. Increased knowledge of our cognitive weaknesses, which is obtained through analyses of vast amounts of personal data, combined with the more powerful means of targeting individuals, based on data on their personalities, proclivities, and weaknesses, makes the techniques of nudging a potent threat to liberty. The third conclusion is that Tocqueville's nightmare, an immaterial tyranny of popular opinion, has resurfaced through our creation of filter bubbles and echo chambers. I discuss how selective exposure, i.e., the human tendency to seek information that supports preestablished beliefs, to avoid the opposite belief, is multiplied by our

Big Data's Threat to Liberty. https://doi.org/10.1016/B978-0-12-823806-6.00003-X
Copyright © 2021 Elsevier Inc. All rights reserved.

use of algorithms, and by the fact that we move toward intragroup homogeneity through echo chambers, while at the same time we get increased polarization and intergroup heterogeneity in society as a whole. In concert, the threats are further exacerbated, as, for example, nudging can be combined with the phenomena discussed as algorithmic curation of information. A key insight derived from the analysis of the three threats is that privacy is key for facing any and all of the threats, and this highlights the reason for emphasizing Big Data's threats to liberty. In sum, the examination of the threats has indicated that they jointly lead to societies in which we lose liberty through surveillance and lack of privacy, lack of alternative lifestyles, manipulation and coercion through the exploitation of the information gathered by surveillance, and a reduction of the breadth of information that allows us to develop into autonomous individuals.

Some of those who have written generically on the threats Big Data poses to liberty implicitly or explicitly suggest that the threats I examine only threaten *positive* liberty and that we must move beyond liberal theory in order to face them. This book has shown, through a wide variety of avenues and perspectives, that such a conclusion is premature.

Liberalism is here taken to be the application of liberal theory in order to foster and develop societies that promote liberty and value pluralism. A key feature of liberal theory is the emphasis on individual liberty and the right of all not to have such rights sacrificed in search for comprehensive conceptions of the good that they themselves might not subscribe to, or approaches that emphasize groups and wholes to such a degree that the individual is compromised. The focus on individualism is one of the reasons some argue that liberal theory must be abandoned. However, I have shown that individualism and liberal theory provide a wide array of reasons to take issue with the problematic aspects of Big Data, and that the protection of privacy, and prevention of manipulation and coercion are crucial concerns for any liberal.

The reasons range from the danger of abuse and exploitation of data to the less obvious ways in which the use of data can influence an individual's actions and selves. There are dangers associated with gathering personal data, as these can be stolen or lost, leading to the potential exploitation and abuse of such data in conflict with the desires and explicit consent of individuals. Furthermore, data can change hands through a myriad of ways that individuals cannot be expected to fully understand, and the durability of data entails risks that future modes of analysis and application have negative effects on individuals that they had not envisaged. Another key reason for liberals to take issue with Big Data is the various ways in which power can be exerted through technologies in ways that constitute interference. Throughout this book I have shown a variety of ways in which nonphysical power can be exerted in ways that interfere with individual liberty and that recognizing this is fully compatible with liberal theory. While the most restrictive interpretations of negative liberty might not be open to considerations of these types of interference, other equally legitimate interpretations do. Furthermore, liberal theory is certainly not limited to considerations of negative liberty alone. As Berlin emphatically emphasized, much controversy surrounding his work stem from the fact that he had omitted certain considerations about the *conditions* of liberty because the thought that they were self-evident—conditions of great importance for any liberal that does not desire to limit themselves to theoretical considerations about what liberty is while disregarding what brings this liberty about.

For liberty to prevail and have real value for individuals, certain conditions must be satisfied. Liberalism is aimed at providing and safeguarding such preconditions. When such conditions cannot be assumed to be created through the spontaneous and free actions of individuals, a liberal condones and encourages government coercion. The condition of liberty that has been most emphasized in this book is privacy. Furthermore, I have argued that privacy is a public good, and this creates a clear imperative for government intervention aimed at resolving the *tragedy of the privacy common*.

Liberal theory allows us to understand the threats posed by Big Data, and it also provides reasons for government to intervene in order to protect individuals from these threats. However, an important facet of the phenomena here examined is that the source of the threats is somewhat different from the threats liberalism has historically concerned itself with. One reason is that government is now not the prime, or at least not the only, source of the threat. Private companies are key actors involved in creating the threats of Big Data, and liberal theory thus provides a justification for government regulation and limitations on the rights of individuals and private companies. However, basing policy on liberal theory simultaneously means that we will in no way be blind to the potential threat from government, and the need to ensure that any strengthening of government powers does not exceed that which is strictly necessary for protecting the rights of individuals.

References

Baum, B., & Nichols, R. (2013). *Isaiah Berlin and the politics of freedom: 'Two concepts of liberty' 50 years later.* New York: Routledge.

MacGilvray, E. (2013). Republicanism and the market in "two concepts of liberty". In B. Baum, & R. Nichols (Eds.), *Isaiah Berlin and the politics of freedom: 'Two concepts of liberty' 50 years later* (Vol. 50, pp. 114–126). New York: Routledge.

Index

A

Absolute negative liberty, 24–25
Active observation, 42
Active surveillance, threat of, 43
Affinity profiling, 28
Algorithms, 19–23, 79–80, 83–85
AlphaGo, 20
AlphaStar, 20
Analysis of concepts, 6
Analytical approach, 16
Analytical philosophy (AP), 5–6
Artificial intelligence (AI), 3–4, 19–23
Artificial neural networks, 22–23
Attribution of action, 114–117
Autonomy, 122
  liberty as, 29–30

B

Behavioral confirmation theory, 101
Behavioural surplus, 18
Big Brother, 38–46
  from Big Data to, 37–38
Big Data, 1, 16–19, 35, 55–58, 134–135
  benefits, 1
  boundaries, 3–5
  and information, 80–87
  as logic and a system of technologies, 15–23
  nudging as threat to liberty, 74–76
  rise of, 54–55
  and superficial voluntariness, 39–40
  surveillance, 35–38
    from Big Data to Big Brother, 37–38

gathering, 35–36
  as threat to liberty, 46–48
Big nudge, 56
Brainwashing, 28

C

Causal responsibility approach, 25, 118
Citizen Score, 42
Coercion, 64–66, 89, 115, 123–126
Cognitive dissonance, 82
Common good, 105
Communalism. See Communism
Communality. See Communism
Communism, 9
Computing Machinery and Intelligence, 20
Conceptual analysis, 7
Confirmation bias, 81–82
Constitutive power, 117
Coordination problem, 105–106
Counterfactuals, 7
Curation of information, 83–85

D

Danger of abuse, 44–45
Deep learning, 22–23
DeepMind, 20
Direct surveillance, 41
Disinterestedness, 9
Domination, 29

E

Echo chambers, 11, 86–87, 93
Emancipation, 26
Endowment effect, 53
Episodic power, 115–116
Expectancy effect, 101–102
Expectations, 79–80, 82

Explicit personalization, 81–82
Externalities, 106

F

Filter bubbles, 11, 21–22, 79, 83–85
Fourth industrial revolution, 2
Framing effects, 53
Freedom, 6, 41–46, 87–91, 133
  under gaze of Big Data, 46–48
  information and individuality, 90–91
  liberty and tyranny as political and social phenomena, 87–90

G

General intelligence, 21
Google, Apple, Facebook, and Amazon (GAFA), 18
Government, 135

H

High-level machine intelligence (HLMI), 21
Human nature, 81–82

I

Idealism, 52
Identity, 55
Implicit personalization, 81–82
Independence, 123
  liberty as, 29–30
Indirect surveillance, 41
Individuality, 90–91
Inertia, 53
Information, 80–87, 90–91
Interference, 122–128
  attributing, 117–118
Intergroup heterogeneity, 85–87

138 Index

Intragroup homogeneity, 85–87
Irrationality, 52–53

L

Liberal democracy, 4–5
Liberal manipulation critique of
 nudging, 56
Liberal policy, implications for,
 128–130
Liberal theory, 113–114,
 133–135
Liberalism, 128, 134–135
Libertarian paternalism, 52
Liberty, 1, 8, 79–80, 113,
 133
 in guises, 23–30
 as nondomination,
  independence, and
  autonomy, 29–30
 nudging and, 58–74
 as political and social
  phenomena, 87–90
 preliminaries, 114–121

M

Machine learning, 19–23
Manipulation, 66–68, 115,
 123–126
Market failure, 105
Methodological foundations,
 5–8
Micro-power, 125–126

N

Negative liberty, 24–26, 59–61,
 120
 threat to, 47–48
Neoliberalism, 125
Nondomination, liberty as,
 29–30
Nonphysical force, 68–71
Nudge, 51, 74–76
 marketing, 58
 powered by Big Data, 54–58
Nudging, 1, 51–54, 57
 and coercion, 64–66

by curation of information,
 100–104
 expectancy effect, 101–102
 Proteus effect, 102–103
 Thomas theorem, 103–104
and liberty, 58–74
theory, 54
tolerating, 71–74

O

Observation, 41
Organized skepticism, 9
Overall freedom, 23–24

P

Passive observation, 41
Passive surveillance, threat of,
 42–43
Paternalism, 61
Perfectionism, 119–121
Perfectionist liberty, 120
Personalization, 81–82
Persuasion, 115
Philosophical analysis, 6
Philosophical foundations,
 5–8
Philosophy, 5–6
Policy, 135
Political philosophy, 5
Political science, 5–6
Positive liberty, 26–30, 61–64,
 134
 threat to, 46–47
Power, 6, 114–117
Privacy, 11, 29, 38–41, 105–109,
 126–128, 133–134
 difficulty of preserving,
  108–109
 general profiles, 107–109
 as precondition, 127–128
 relational leakages, 106–107
Private information, 38–41
Propositions, 7
Proteus effect, 102–103
Psychic coercion, 89
Psychological coercion, 68

Psychological force, 26, 68
Psychological manipulation, 26
Public goods, 105, 108, 135

R

Radical empiricism, 103–104
Ranking of values, 119–121
Rational persuasion, 51, 59, 66
Reinforcement learning, 22–23
Relational leakages, 106–107
Rent seeking behavior, 109
Republican liberty, 29–30

S

Secret coercion, 66–68
Selective exposure, 79, 81–82,
 133–134
Self-abnegation, 27
Self-mastery, 26
Self-realization, 27
Singularity, 21
Soft power, 116–117
Superficial voluntariness,
 39–40
Superintelligence, 21
Surveillance, 1–2, 41–46, 99,
 133–134
 capitalism, 18, 80, 130
 threat of surveillance proper,
  43–44
Surveillant assemblage, 35–36,
 54–55
Systemic power, 116–117

T

Technology, 3
 Big Data as logic and system
  of, 15–23
Theoretical foundations, 5–8
Thomas theorem, 103–104
Threat(s), 1, 15, 99, 123–126
 of active surveillance, 43
 to negative liberty, 47–48
 of passive surveillance, 42–43
 to positive liberty, 46–47
 of surveillance proper, 43–44

of tyranny of perceived opinion, 92–95

propositions into premises, 92–93

Tragedy of the commons, 106

Training sets, 22

Transparency, 55

Turing test, 20

Tyranny

of perceived opinion, 93–95

as political and social phenomena, 87–90

U

Ultraintelligence, 21

Universalism, 9

V

Veil of complexity, 117–118

Volume, velocity, and variety (three Vs), 17

Z

Zuboff's surveillance capitalism, 80–81

Printed in the United States
by Baker & Taylor Publisher Services